INTEGRAL
TRANSFORMS
IN APPLIED
MATHEMATICS

INTEGRAL TRANSFORMS IN APPLIED MATHEMATICS

John W. Miles

UNIVERSITY OF CALIFORNIA, SAN DIEGO

CAMBRIDGE

AT THE UNIVERSITY PRESS · 1971

CAMBRIDGE UNIVERSITY PRESS
Cambridge, New York, Melbourne, Madrid, Cape Town, Singapore, São Paulo, Delhi

Cambridge University Press
The Edinburgh Building, Cambridge CB2 8RU, UK

Published in the United States of America by Cambridge University Press, New York

www.cambridge.org
Information on this title: www.cambridge.org/9780521083744

First published 1971
This digitally printed version 2008

A catalogue record for this publication is available from the British Library

Library of Congress Catalogue Card Number: 70-172834

ISBN 978-0-521-08374-4 hardback
ISBN 978-0-521-09068-1 paperback

PREFACE

The following treatment of integral transforms in applied mathematics is directed primarily toward senior and graduate students in engineering and applied science. It assumes a basic knowledge of complex variables and contour integration, gamma and Bessel functions, partial differential equations, and continuum mechanics. Examples and exercises are drawn from the fields of electric circuits, mechanical vibration and wave motion, heat conduction, and fluid mechanics. It is not essential that the student have a detailed familiarity with all of these fields, but knowledge of at least some of them is important for motivation (terms that may be unfamiliar to the student are listed in the *Glossary*, p. 89). The unstarred exercises, including those posed parenthetically in the text, form an integral part of the treatment; the starred exercises and sections are rather more difficult than those that are unstarred.

I have found that all of the material, plus supplementary material on asymptotic methods, can be covered in a single quarter by first-year graduate students (the minimum preparation of these students includes the equivalent of one-quarter courses on each of complex variables and partial differential equations); a semester allows either a separate treatment of contour integration or a more thorough treatment of asymptotic methods. The material in Chapter 4 and Sections 5.5 through 5.7 could be omitted in an undergraduate course for students with an inadequate knowledge of Bessel functions.

The exercises and, with a few exceptions, the examples require only those transform pairs listed in the *Tables* in Appendix 2. It is scarcely necessary to add, however, that the effective use of integral transforms in applied mathematics eventually requires familiarity with more extended tables, such as those of Erdélyi, Magnus, Oberhettinger, and Tricomi (herein abbreviated EMOT, followed by the appropriate entry number).

Chapter 1 and minor portions of Chapters 2 through 5 are based on a lecture originally given at various points in California in 1958 and subsequently published [Beckenbach (1961)] by McGraw-Hill. I am indebted to the McGraw-Hill Publishing Company for permission to reuse portions of the original lecture; to Cambridge University Press for permission to reproduce Figures 4.2 and 4.3; to Professors D. J. Benney and W. Prager for helpful criticism; to Mrs. Elaine Blackmore for preparation of the typescript and to Mr. Y. J. Desaubies for his aid in reading the proofs.

J. W. M.

La Jolla, 1968

CONTENTS

To Oliver Heaviside

1 INTEGRAL-TRANSFORM PAIRS

1.1 Introduction

We define

$$F(p) = \int_a^b K(p, x) f(x)\, dx \qquad (1.1.1)$$

to be an *integral transform* of the function $f(x)$; $K(p, x)$, a prescribed function of p and x, is the *kernel* of the transform. The introduction of such a transform in a particular problem may be advantageous if the determination or manipulation of $F(p)$ is simpler than that of $f(x)$, much as the introduction of $\log x$ in place of x is advantageous in certain arithmetical operations. The representation of $f(x)$ by $F(p)$ is, in many applications, merely a way of organizing a solution more efficiently, as in the introduction of logarithms for multiplication, but in some instances it affords solutions to otherwise apparently intractable problems, just as in the introduction of logarithms for the extraction of the 137th root of a given number.

Today, the *Laplace transform* $[K = e^{-px}, a = 0,$ and $b = \infty$ in (1.1.1)],

$$F(p) = \int_0^\infty e^{-px} f(x)\, dx, \qquad (1.1.2)\dagger$$

may be claimed as a working tool for the solution of ordinary differential equations by every well-trained engineer. We consider here its application

---

† In Heaviside's form of operational calculus, the right-hand side of (1.1.2) appears multiplied by p, but the form (1.1.2) is now almost universal.

1

to both ordinary and partial differential equations. We also consider applications to partial differential equations of the *Fourier transform* (also called the *exponential-Fourier transform* or the *complex Fourier transform*),

$$F(p) = \int_{-\infty}^{\infty} e^{-ipx} f(x)\, dx, \qquad (1.1.3)$$

the *Fourier-cosine* and *Fourier-sine transforms*,

$$F(p) = \int_{0}^{\infty} f(x) \cos px\, dx \qquad (1.1.4)$$

and

$$F(p) = \int_{0}^{\infty} f(x) \sin px\, dx, \qquad (1.1.5)$$

and the *Hankel transform* (also called the *Bessel* or *Fourier–Bessel transform*),

$$F(p) = \int_{0}^{\infty} f(x) J_n(px) x\, dx. \qquad (1.1.6)$$

Our definitions are those of Erdélyi, Magnus, Oberhettinger, and Tricomi (1954, hereinafter abbreviated as EMOT) except for (1.1.6); other definitions of the Fourier transforms, differing from those of (1.1.3) to (1.1.5) by constant factors, are not uncommon. Notations for the transforms themselves vary widely, and symbols other than p are not uncommon for the arguments. In particular, engineers typically use s, rather than p, in the Laplace transforms of time-dependent functions [in which case t appears in place of x in (1.1.2)]. The properties of the foregoing infinite transforms are summarized in Table 2.3 (Appendix 2, p. 85). We consider these and other properties, together with applications, in the following chapters.

Consider the following elementary example, in which we anticipate certain results that are derived in Chapter 2. We require the charge $q(t)$ on a capacitor C, in series with a resistor R, following the application of a constant voltage v at $t = 0$. The differential equation obtained by equating the sum of the voltages across each of C and R to the applied voltage is

$$R \frac{dq}{dt} + C^{-1} q = v. \qquad (1.1.7)$$

Let

$$Q(p) = \int_{0}^{\infty} e^{-pt} q(t)\, dt$$

be the Laplace transform of $q(t)$. Taking the Laplace transform of (1.1.7), transforming the derivative through integration by parts,

$$\int_0^\infty e^{-pt} \frac{dq}{dt}\, dt = pQ(p) - q(0),$$

and invoking the fact that v is a constant, we obtain

$$(Rp + C^{-1})Q(p) - Rq(0) = vp^{-1}.$$

Solving for $Q(p)$, we obtain

$$Q(p) = \frac{q(0)}{p + (RC)^{-1}} + \frac{(v/R)}{p[p + (RC)^{-1}]},$$

which we rewrite in the form

$$Q(p) = q(0)(p + \alpha)^{-1} + Cv[p^{-1} - (p + \alpha)^{-1}], \qquad (1.1.8)$$

where $\alpha = 1/RC$. Referring to entry 2.1.2, in Table 2.1 (Appendix 2, p. 83), we find that the functions whose transforms are $1/(p + \alpha)$ and $1/p$ are $\exp(-\alpha t)$ and 1, the latter being a special case of the former. Invoking these results in (1.1.8), we obtain

$$q(t) = q(0)e^{-\alpha t} + Cv(1 - e^{-\alpha t}). \qquad (1.1.9)$$

This example is too straightforward to illustrate the real power of the Laplace transform, but it does serve to illustrate such basic advantages as the reduction of differential to algebraic operations and the automatic incorporation of initial conditions. [Bracewell (1965) and Gardner–Barnes (1942) give extensive applications of Fourier and Laplace transforms, respectively, to circuit analysis.]

The transforms defined by (1.1.2) to (1.1.6) are the only infinite ones in widespread use at this time, but many others have been studied and tabulated (see EMOT, Vol. 2), and still others may be introduced in the future. If one's goal is merely to produce formal solutions, it suffices to know the inversion formula that determines $f(x)$ from $F(p)$, but extensive tabulations of $f(x)$ versus $F(p)$ are essential in applied mathematics. Returning to our analog of the logarithm, we note that formal analysis requires only the knowledge that the inverse of $y = \log_a x$ is $x = a^y$, whereas a table of x versus y is indispensable for numerical computation.

In addition to the possibility of defining new transforms through new kernels, there is also the possibility of adopting finite limits in (1.1.1), thereby obtaining so-called *finite transforms*. If, for example, we replace the upper limits in (1.1.4) and (1.1.5) by π, the inversion formulas are ordinary Fourier-cosine and -sine series summed over integral values of p. More generally, if the kernel $K(p, x)$ in (1.1.1) yields a set of functions orthogonal, with suitable weighting function, over the interval a, b for an infinite discrete set of values p, then the inversion formula defines a Fourier-type series.

The result of introducing a finite Fourier transform in a given problem is merely to mechanize the classical technique of Fourier series; however, it is generally true that the more tedious solution of the problem by the classical technique is straightforward, albeit sometimes calling for greater ingenuity (compare the use of Lagrange's equations in mechanics). This is to be contrasted with the applications of infinite transforms, which frequently offer entirely new insight and reduce transcendental to algebraic operations, thereby affording solutions to problems that might have required far greater ingenuity for their solution by classical techniques. Nevertheless, the student must bear in mind that, with few exceptions, integral transforms are applicable only to linear differential and/or integral equations with either constant coefficients or (as with the Hankel transform) very special nonconstant coefficients.

Integral transforms in applied mathematics find their antecedents in the classical methods of Fourier and in the operational methods of Heaviside, antecedents that had rather different receptions by contemporary mathematicians. The eleventh edition (1910) of the *Encyclopedia Britannica* devotes five pages to Fourier series but does not mention Heaviside's operational calculus; indeed, no direct entry appears for Heaviside in that edition, although his name is mentioned peripherally. The fourteenth edition (1942) does contain a brief biographical entry on Heaviside, but the only reference to his operational calculus is the rather oblique statement that "he made use of unusual methods of his own in solving his problems."

Fourier's theorem has constituted one of the cornerstones of mathematical physics from the publication of his *La Théorie Analytique de la Chaleur* (1822), and its importance was quickly appreciated by mathematicians and physicists alike. For example, Thomson and Tait remarked that

... Fourier's Theorem ... is not only one of the most beautiful results of modern analysis, but may be said to furnish an indispensable instrument in the treatment of nearly every recondite question in modern physics. To mention only sonorous vibrations, the propagation of electrical signals along a telegraph wire, and the conduction of heat by the earth's crust, as subjects in their generality intractable without it, is to give but a feeble idea of its importance.

The concept of an integral transform follows directly from Fourier's theorem (see Section 1.2 below), but the historical approach, at least to physical applications, was largely through operational methods. Operational methods in mathematical analysis, having been introduced originally by Leibniz, are nearly as old as the calculus, but their widespread use in modern technology stems almost entirely from the solitary genius of Oliver Heaviside (1850–1925). To be sure, the bases of Heaviside's method, as he recognized and stated, lay in the earlier work of Laplace and Cauchy, but it was Heaviside who recognized and exposed the power of operational methods not only in circuit analysis but also in partial differential equations. Unlike Fourier, Heaviside had no university training and was not a recognized mathematician; indeed, he scorned not only mathematical rigor ("Shall I refuse my dinner because I do not fully understand the process of digestion?") but also, it sometimes appeared, mathematicians ("Even Cambridge mathematicians deserve justice."). This lack of rapport with mathematicians may have delayed the full appreciation of his work. Even some modern mathematicians have been reluctant to give Heaviside his due; thus, Van der Pol–Bremmer (1950) criticize Doetsch (1943) for his description of Heaviside as merely "ein englischer Elektroingenieur," using methods that were "mathematisch sehr unzulänglich" and "allerdings mathematisch unzureichend." E. T. Whittaker, on the other hand, offered the following evaluation (in Heaviside's obituary): "Looking back . . . , we should now place the operational calculus with Poincaré's discovery of automorphic functions and Ricci's discovery of the tensor calculus as the three most important mathematical advances of the last quarter of the nineteenth century."

We conclude this introduction by contrasting the approaches of the pure mathematician and the physical scientist to transform theory. At one extreme we have Titchmarsh's statement, in the preface to his treatise (1948), that "I have retained, as having a certain picturesqueness, some references to 'heat,' 'radiation,' and so forth; but the interest is purely analytical, and the reader need not know whether such things exist." At the other, we have Heaviside's cavalier statement, "The mathematicians say this series diverges; therefore, it should be useful." In the following presentation of integral-transform theory, we attempt to follow the line set down by Lord Rayleigh:

> In the mathematical investigation I have usually employed such methods as present themselves naturally to a physicist. The pure mathematician will complain, and (it must be confessed) sometimes with justice, of deficient rigour. But to this question there are two sides. For, however important it may be to maintain a uniformly high standard in pure mathematics, the physicist may occasionally do well to rest content with arguments which are fairly satisfactory

and conclusive from his point of view. To his mind, exercised in a different order of ideas, the more severe procedure of the pure mathematician may appear not more but less demonstrative. And further, in many cases of difficulty to insist upon the highest standard would mean the exclusion of the subject altogether in view of the space that would be required.

1.2 Fourier's integral formulas

We first give a formal derivation of Fourier's theorem in complex form, following in all essential respects the argument offered by Fourier himself.† Let $f(x)$ be represented by the complex Fourier series

$$f(x) = \sum_{n=-\infty}^{\infty} c_n \exp(ik_n x) \qquad (-\tfrac{1}{2}\lambda < x < \tfrac{1}{2}\lambda), \qquad (1.2.1)$$

where

$$c_n = \lambda^{-1} \int_{-\lambda/2}^{\lambda/2} f(\xi) \exp(-ik_n \xi) \, d\xi \qquad (1.2.2)$$

and

$$k_n = \frac{2n\pi}{\lambda}. \qquad (1.2.3)$$

This representation is evidently periodic with a wavelength λ. We now allow λ to tend to infinity, noting that the consecutive k_n are separated by the increment $\Delta k = 2\pi/\lambda$; then, combining (1.2.1) and (1.2.2), we obtain

$$f(x) = \lim_{\lambda \to \infty} \sum_{n=-\infty}^{\infty} \exp(ik_n x) \frac{\Delta k}{2\pi} \int_{-\lambda/2}^{\lambda/2} f(\xi) \exp(-ik_n \xi) \, d\xi. \qquad (1.2.4)$$

Replacing the sum by an integral in the limit, we obtain

$$f(x) = \frac{1}{2\pi} \int_{-\infty}^{\infty} dk \int_{-\infty}^{\infty} f(\xi) \exp[ik(x-\xi)] \, d\xi. \qquad (1.2.5)$$

Expressing the exponential in terms of its trigonometric components and invoking the even and odd nature of $\cos k(x-\xi)$ and $\sin k(x-\xi)$, respectively, as functions of k, we obtain *Fourier's integral formula*:

$$f(x) = \frac{1}{\pi} \int_{0}^{\infty} dk \int_{-\infty}^{\infty} f(\xi) \cos k(x-\xi) \, d\xi. \qquad (1.2.6)$$

Fourier's derivation differed from the above only in starting from the trigonometric form of his series. We emphasize that the order of integration in (1.2.5) and (1.2.6) must be preserved; its reversal would lead to

---

†This derivation is included in order to emphasize the relation between Fourier series and Fourier integrals. It is assumed that the student has had previous experience with both Fourier series and Fourier integrals and that he has been exposed to a more rigorous derivation of Fourier's integral theorem [see, e.g., Churchill (1963)].

meaningless integrals. On the other hand, we assume that $f(x)$ vanishes with sufficient rapidity for large $|x|$ to ensure the existence of the double integrals as written. Actually, f vanishes exponentially in typical, not-too-idealized, physical problems.

Now let $f(x)$ be either an even or an odd function (any function that is not even or odd can be split into a sum of two such functions). Expanding the cosine in (1.2.6), we obtain *Fourier's cosine formula*,

$$f(x) = f(-x) = \frac{2}{\pi} \int_0^\infty \cos kx \, dk \int_0^\infty f(\xi) \cos k\xi \, d\xi \qquad (1.2.7)$$

for an even function, and *Fourier's sine formula*,

$$f(x) = -f(-x) = \frac{2}{\pi} \int_0^\infty \sin kx \, dk \int_0^\infty f(\xi) \sin k\xi \, d\xi \qquad (1.2.8)$$

for an odd function.

1.3 Fourier-transform pairs

Equation (1.2.5) may be resolved into the transform pair

$$F(k) = \mathscr{F}\{f\} \equiv \int_{-\infty}^\infty f(x)e^{-ikx} \, dx \qquad (1.3.1a)$$

and

$$f(x) = \mathscr{F}^{-1}\{F\} \equiv \frac{1}{2\pi} \int_{-\infty}^\infty F(k)e^{ikx} \, dk, \qquad (1.3.1b)$$

where $\mathscr{F}$ is the Fourier-transform operator and $\mathscr{F}^{-1}$ is its inverse (we omit the braces around the operand when this can be done without ambiguity). The location of the factor $1/(2\pi)$ is essentially arbitrary; from the viewpoint of establishing an analogy between Fourier series and Fourier integrals, it would appear preferable to place it in (1.3.1a) rather than (1.3.1b), whereas from an aesthetic viewpoint a symmetric disposal of the identical factors $(2\pi)^{-1/2}$ would be desirable. Each of these conventions has been adopted by various writers, but the form chosen in (1.3.1a, b) has two major advantages. First, it agrees with the notation adopted by Campbell–Foster (1948), in their very extensive table of Fourier transforms (see also EMOT, Vol. 2). Secondly, it affords a direct transition to the accepted definition of the Laplace-transform pair (see Section 1.4). A third advantage, of special interest in electric-circuit or wave-motion problems, is that if x, implicitly a space variable in the foregoing discussion, is replaced by the time-variable t and k is replaced by $2\pi v$, where v is a frequency, (1.3.1a, b) go over to the symmetric pair

$$F(v) = \int_{-\infty}^{\infty} f(t) e^{-2\pi i v t} \, dt \qquad (1.3.2a)$$

and

$$f(t) = \int_{-\infty}^{\infty} F(v) e^{2\pi i v t} \, dv, \qquad (1.3.2b)$$

in which $f(t)$ is represented as a spectral superposition of simple harmonic oscillations of frequency v and complex amplitude $F(v)$. We remark that a similar form for the space-variable pair of (1.3.1a, b) follows from the substitution $k = 2\pi\kappa$, where κ is a reciprocal wavelength.

Consider, for example,

$$f(x) = e^{-a|x|} \qquad (-\infty < x < \infty),$$

the substitution of which into (1.3.1a) yields

$$\begin{aligned}
\mathscr{F}\{e^{-a|x|}\} &= \int_{-\infty}^{\infty} e^{-a|x| - ikx} \, dx \\
&= \int_{0}^{\infty} e^{-(a+ik)x} \, dx + \int_{-\infty}^{0} e^{(a-ik)x} \, dx \\
&= (a + ik)^{-1} + (a - ik)^{-1} \\
&= 2a(a^2 + k^2)^{-1}.
\end{aligned} \qquad (1.3.3)$$

Invoking (1.3.1b), closing the path of integration (see Figure 1.1) by a semi-circle of infinite radius in $k_i > 0$† for $x > 0$ or in $k_i < 0$ for $x < 0$, observing that the integrand has poles at $k = \pm ia$, and invoking Cauchy's residue theorem, we obtain

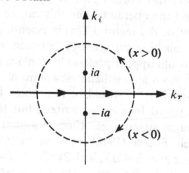

FIGURE 1.1   Path(s) of integration and poles for the example of (1.3.4). The semicircles have radii that tend to infinity.

---

† We use the subscripts r and i here and subsequently to designate the real and imaginary parts of a complex variable.

$$\mathscr{F}^{-1}\{2a(a^2 + k^2)^{-1}\} = \frac{a}{\pi} \int_{-\infty}^{\infty} (a^2 + k^2)^{-1} e^{ikx} \, dk$$

$$= \frac{a}{\pi} (\pm 2\pi i) R\{(a^2 + k^2)^{-1} e^{ikx}\}|_{k = \pm ia} \qquad (x \gtrless 0)$$

$$= \frac{a}{\pi} (\pm 2\pi i)(\pm 2ia)^{-1} e^{\mp ax} \qquad (x \gtrless 0)$$

$$= e^{-a|x|}, \qquad (1.3.4)$$

where the upper and lower signs correspond to $x > 0$ and $x < 0$, respectively, and $R\{\ \}$ is the residue of the bracketed quantity.

Fourier's cosine and sine formulas, (1.2.7) and (1.2.8), may be resolved into the transform pairs

$$F(k) = \mathscr{F}_c\{f\} \equiv \int_0^{\infty} f(x) \cos kx \, dx, \qquad (1.3.5a)$$

$$f(x) = \mathscr{F}_c^{-1}\{F\} \equiv \frac{2}{\pi} \int_0^{\infty} F(k) \cos kx \, dk \qquad (1.3.5b)$$

and

$$F(k) = \mathscr{F}_s\{f\} \equiv \int_0^{\infty} f(x) \sin kx \, dx, \qquad (1.3.6a)$$

$$f(x) = \mathscr{F}_s^{-1}\{F\} \equiv \frac{2}{\pi} \int_0^{\infty} F(k) \sin kx \, dk, \qquad (1.3.6b)$$

thereby providing the inversions of (1.1.4) and (1.1.5), respectively. Again, the $2/\pi$ factors may be resolved differently—in particular, symmetrically; the notation adopted here is that of EMOT. We remark that the Fourier-cosine (-sine) transform is suited either to a function that is defined only in $x > 0$ or to an even (odd) function of x; conversely, the function defined by (1.3.5b) or (1.3.6b) is an even or odd function, respectively, of x.

Consider, for example,

$$f(x) = e^{-ax} \qquad (a > 0,\ 0 \leqq x < \infty),$$

the substitution of which into (1.3.5a) yields

$$\mathscr{F}_c\{e^{-ax}\} = \int_0^{\infty} e^{-ax} \cos kx \, dx$$

$$= \tfrac{1}{2} \int_0^{\infty} e^{-ax}(e^{ikx} + e^{-ikx}) \, dx$$

$$= \tfrac{1}{2}[(a - ik)^{-1} + (a + ik)^{-1}]$$

$$= a(a^2 + k^2)^{-1}. \qquad (1.3.7)$$

We remark that (1.3.7) follows almost directly from the example of (1.3.3) (or conversely) by virtue of the fact that $\exp(-a|x|)$ is an even function of x. The student should show that

$$\mathscr{F}_s\{e^{-ax}\} = k(a^2 + k^2)^{-1} \qquad (1.3.8)$$

and observe that this result is *not* directly related to the example of (1.3.3) (see the last sentence in the preceding paragraph).

1.4 Laplace-transform pairs

The path of integration for (1.3.1b) may be deformed into a complex-k plane in any manner that ensures the convergence of the integrals for both $F(k)$ and $f(x)$. Suppose, to cite the most important, special case, that $\exp(-cx)f(x)$ vanishes appropriately at both limits; in particular, $f(x)$ may vanish identically in $x < 0$. Then the modified transform

$$F(k) = \mathscr{F}\{f\} = \int_{-\infty}^{\infty} f(x)e^{-ikx}\,dx \qquad (k_i = -c) \qquad (1.4.1a)$$

exists, where k_i denotes the imaginary part of k, and $f(x)$ is given by

$$f(x) = \mathscr{F}^{-1}\{F\} = \frac{1}{2\pi} \int_{-\infty-ic}^{\infty-ic} F(k)e^{ikx}\,dk. \qquad (1.4.1b)$$

Thus, Fourier's integral formula extends to functions for which (1.2.5) might not be valid. In the most important, physical applications, c is positive and the path of integration appears as in Figure 1.2a.

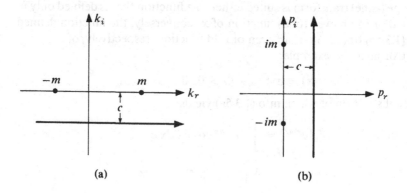

(a)   (b)

FIGURE 1.2   Paths of integration for the inverse transforms of (1.4.1b) and (1.4.3b) and the poles for the corresponding transforms of sin mx.

If we now rotate the path of integration through a right angle (see Figure 1.2b), introduce $p = ik$, and at the same time replace $F(k)$ by $F(p)$, (1.4.1a) and (1.4.1b) go over to the *two-sided Laplace-transform pair*,†

$$F(p) = \int_{-\infty}^{\infty} f(x)e^{-px}\,dx \qquad (p_r = c) \qquad (1.4.2a)$$

and

$$f(x) = \frac{1}{2\pi i}\int_{c-i\infty}^{c+i\infty} F(p)e^{px}\,dp. \qquad (1.4.2b)$$

Finally, we suppose that $f(x)$ vanishes for $x < 0$, so that the lower limit in (1.4.2a) may be replaced by zero. The resulting integral converges if the real part of p exceeds some minimum value, say c_*, not necessarily nonnegative, such that all singularities of $F(p)$ lie in $p_r < c_*$, where p_r denotes the real part of p; the inverse transform (1.4.2b) exists for all $c > c_*$, and we have the Laplace-transform pair

$$F(p) = \mathscr{L}\{f(x)\} \equiv \int_{0}^{\infty} e^{-px}f(x)\,dx \qquad (p_r > c_*) \qquad (1.4.3a)$$

and

$$f(x) = \mathscr{L}^{-1}\{F(p)\} \equiv \frac{1}{2\pi i}\int_{c-i\infty}^{c+i\infty} F(p)e^{px}\,dp \qquad (c > c_*). \qquad (1.4.3b)$$

Consider, for example,

$$f(x) = \begin{cases} \sin mx \\ 0 \end{cases} \qquad (x \gtrless 0), \qquad (1.4.4)$$

for which the Fourier transform defined by (1.3.1a) does not exist. Invoking (1.4.1a), we obtain (the student should fill in the details)

$$\mathscr{F}\{\sin mx\} = m(m^2 - k^2)^{-1} \qquad (k_i < 0), \qquad (1.4.5)$$

which has poles at $k = \pm m$ on the real axis (see Figure 1.2a) but is an analytic function of k in $k_i < 0$. Similarly,

$$\mathscr{L}\{\sin mx\} = m(m^2 + p^2)^{-1} \qquad (p_r > 0). \qquad (1.4.6)$$

The student should evaluate the corresponding inverse transforms, with the aid of the calculus of residues, to recover the original function.

---

†Applications of the two-sided Laplace transform are considered by Van der Pol–Bremmer (1950).

EXERCISES

1.1 Derive the *Mellin-transform* pair from the Laplace-transform pair of (1.4.2a, b):

$$F(p) = \int_0^\infty x^{p-1} f(x) \, dx \qquad (p_r > c_*)$$

and

$$f(x) = \frac{1}{2\pi i} \int_{c-i\infty}^{c+i\infty} F(p) x^{-p} \, dp \qquad (c > c_*).$$

1.2 Derive (1.3.8).

2 THE LAPLACE TRANSFORM

2.1 Introduction

The Laplace transform, it can be fairly said, stands first in importance among all integral transforms; for, while there are many specific examples in which other transforms prove more expedient, the Laplace transform is both the most powerful in dealing with initial-value problems and the most extensively tabulated. We consider in this chapter some of the fundamental properties that give it this flexibility and illustrate these properties by application to typical problems in vibrations, wave propagation, heat conduction, and aerodynamics.

The Laplace-transform pairs required in this chapter are tabulated in Table 2.1 (Appendix 2, p. 83). These results may be extended in various ways with the aid of the operational formulas of Table 2.2 (Appendix 2, p. 84). We give derivations of the more important formulas in the text and set the derivations of others as exercises. We refer to entries in these tables by the prefix T, followed by the appropriate entry number. More extensive tables are listed under Bibliography. We regard Erdélyi *et al.* (EMOT) as a standard reference, although the most extensive table of Laplace transforms now available is that of Roberts–Kaufman (1966).

Many of the Laplace transforms that arise in the analysis of a dynamical system with a finite number of degrees of freedom have the form

$$F(p) = \frac{G(p)}{H(p)}, \tag{2.1.1}$$

where G and H are polynomials, of degree M and N, $N > M$. The inversion of $F(p)$ may be inferred from a suitable partial-fraction expansion, together with the basic entries of Table 2.1. The basic techniques for partial-fraction expansions are developed in Appendix 1. A general expansion, subject to the restriction that the zeros of $H(p)$ be simple, is given by (2.7.7) (see also Exercise 2.8). It may be more direct, in many applications, to obtain appropriate, partial-fraction expansions by inspection, as in the examples of (2.3.7), (2.6.13), and (2.6.16).

The general procedure for the application of an integral transform to boundary- and/or initial-value problems comprises the following steps:

(a) Select and apply a transform that is appropriate to the partial differential equation† and its boundary and/or initial conditions in the sense that it reduces the differential operations with respect to a given independent variable to algebraic operations. Different transformations may be applied simultaneously to different variables—e.g., a Laplace transformation with respect to time and a Hankel transform with respect to a radial variable, as in Section 4.3.

(b) Solve the transformed problem for the transform.

(c) Invert the transform through any or all of the following procedures: (i) preliminary simplification through operational theorems, (ii) direct use of tables, (iii) direct use of the inversion integral, (iv) approximate inversions through appropriate expansions of the transform or modifications of the contour integral. Asymptotic approximations such as may be obtained through the invocation of Watson's lemma (Section 2.7) or the method of stationary phase (Section 3.7) are especially important.

1.2 Transforms of derivatives

The Laplace transformation is typically applied to initial-value problems in order to reduce the derivatives of some function, say $f(t)$, to algebraic form. Let

$$F(p) = \mathscr{L}\{f(t)\} \equiv \int_0^\infty e^{-pt} f(t)\, dt \qquad (2.2.1)$$

denote the Laplace transform of $f(t)$.‡ We obtain the transforms of the first and second derivatives, $f'(t)$ and $f''(t)$, through integration by parts (the student should fill in the details of these integrations):

---

† Integral transforms also may be useful for the solution of integral equations, but we do not consider such applications here. See Sneddon (1951), Tranter (1956), and Van der Pol–Bremmer (1950).

‡ We use t as the independent variable in most of this chapter in recognition of the fact that usually that variable is time; nevertheless, initial-value problems may be encountered in which t is a space variable—e.g., those in linearized, supersonic flow, as in Section 2.10

$$\mathscr{L}\{f'(t)\} = pF(p) - f(0) \qquad (2.2.2a)$$

and
$$\mathscr{L}\{f''(t)\} = p^2F(p) - pf(0) - f'(0), \qquad (2.2.2b)$$

where $f'(0)$ denotes the initial $(t \to 0+)$ value of $f'(t)$. Similarly, we obtain the transform of the nth derivative. $f^{(n)}(t)$, by integrating n times by parts:

$$\mathscr{L}\{f^{(n)}(t)\} = p^nF(p) - \sum_{m=0}^{n-1} p^{n-m-1}f^{(m)}(0), \qquad (2.2.3)$$

where
$$f^{(m)}(0) \equiv \lim_{t \to 0+} \frac{d^m f(t)}{dt^m}. \qquad (2.2.4)$$

One of the virtues of the Laplace transformation for initial-value problems is the systematic manner in which the initial values, $f(0), \ldots,$ $f^{(n-1)}(0)$, are incorporated in the calculation. If, on the other hand, one (or more) of these initial values is not specified, it must be regarded as an unknown parameter, to be determined by some other condition; for example, $f'(0)$ might have to be determined in such a way as to ensure that $f(t)$ has a prescribed value at $t = t_1$, and the Laplace transform then will be less efficient than in the solution of the corresponding initial-value problem.

There exists an important class of problems, dealing with systems that are initially at rest, in which each of the initial values, $f(0), \ldots,$ $f^{(n-1)}(0)$ is zero; then, and only then, we have the Heaviside operational rule

$$\mathscr{L}^{-1}\{p^nF(p)\} = f^{(n)}(t) \qquad [f(0) = \cdots = f^{(n-1)}(0) \equiv 0]. \qquad (2.2.5)$$

A related but generally valid, result is

$$\mathscr{L}^{-1}\{p^{-n}F(p)\} = \int_0^t \cdots \int_0^t f(t)(dt)^n. \qquad (2.2.6)$$

2.3 Simple oscillator

We consider (Figure 2.1) a one-dimensional oscillator of mass m and spring constant k (such that a displacement x is opposed by a restoring force $-kx$) that is subjected to a constant force w_0 in the positive-x direction at $t = 0$. Given the initial displacement and velocity of m, x_0 and

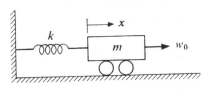

FIGURE 2.1   The oscillator of Section 2.3.

v_0, we seek $x(t)$. [The student presumably will be familiar with the non-operational treatment of this elementary example. We introduce it, together with its subsequent modifications, in order to illustrate basic operational procedures. Gardner–Barnes (1942) and Thomson (1960) give extensive applications of the Laplace transform to more sophisticated mechanical-vibration problems.]

The equation of motion, obtained by applying Newton's second law to the mass m under the action of the forces w_0 and $-kx$, is

$$mx''(t) + kx(t) = w_0. \qquad (2.3.1)$$

The initial conditions are

$$x(0) = x_0 \quad \text{and} \quad x'(0) = v_0. \qquad (2.3.2)$$

Let $X(p)$ denote the Laplace transform of $x(t)$. Transforming the left-hand side of (2.3.1) through (2.2.2b) and w_0 through T2.1.1, we obtain

$$m(p^2 X - px_0 - v_0) + kX = w_0 p^{-1}. \qquad (2.3.3)$$

Solving (2.3.3) for X and introducing the natural frequency of the oscillator,

$$\beta = \left(\frac{k}{m}\right)^{1/2}. \qquad (2.3.4)$$

we obtain

$$X(p) = (p^2 + \beta^2)^{-1}\left[x_0 p + v_0 + \frac{w_0}{m}p^{-1}\right]. \qquad (2.3.5)$$

The inverse transforms of the terms in x_0 and v_0 on the right-hand side of (2.3.5) are given by T2.1.3 and T2.1.4. The remaining term does not appear in Table 2.1, but we may invert it either by applying T2.2.4 to T2.1.4 according to

$$\mathscr{L}^{-1}\{p^{-1}(p^2 + \beta^2)^{-1}\} = \beta^{-1}\int_0^t \sin \beta t\, dt = \beta^{-2}(1 - \cos \beta t) \quad (2.3.6)$$

or by invoking the partial-fraction expansion†

$$\frac{1}{p(p^2 + \beta^2)} = \frac{1}{\beta^2}\left(\frac{1}{p} - \frac{p}{p^2 + \beta^2}\right), \qquad (2.3.7)$$

which yields the same result. Invoking T2.1.3, T2.1.4, and (2.3.6) in (2.3.5) and setting $m\beta^2 = k$, we obtain

$$x(t) = x_0 \cos \beta t + \frac{v_0}{\beta}\sin \beta t + \frac{w_0}{k}(1 - \cos \beta t). \qquad (2.3.8)$$

---

†The student will find, with some experience, that simple partial-fraction expansions may be constructed by inspection. Formal rules for partial-fraction expansions are given in Appendix 1, where the derivation of (2.3.7) is given as an example.

The term w_0/k on the right-hand side of (2.3.8) represents the static displacement of the oscillator under the force w_0; the remaining terms represent free oscillations, which would have been transient (decaying as $t \rightarrow \infty$) if the oscillator had been damped. The first two terms on the right-hand side of (2.3.8) represent the response of the oscillator to the initial displacement and velocity and would have the same form for any applied force. The third term represents the response to the applied force from an initial state of rest (both the displacement and the velocity given by the third term vanish at $t = 0$). Replacing w_0 on the right-hand side of (2.3.1) by some other force, $w(t)$, alters the third term but not the first and second; accordingly, we take $x_0 = v_0 = 0$ in the subsequent variations on this example.

2.4 Convolution theorem

It frequently proves expedient to resolve a Laplace transform into a product of two transforms, either because the inversions of the latter transforms are known or because one of them represents an arbitrary function—typically an input to some physical system. Let $F_1(p)$ and $F_2(p)$ be the Laplace transforms of $f_1(t)$ and $f_2(t)$; the *convolution theorem* states that

$$\mathscr{L}^{-1}\{F_1(p)F_2(p)\} = \int_0^t f_1(t - \tau)f_2(\tau)\,d\tau. \qquad (2.4.1)$$

The convolution integral on the right-hand side of (2.4.1) is often denoted by $f_1(t) * f_2(t)$.

To prove (2.4.1), we form the product of the defining integrals for F_1 and F_2 to obtain

$$F_1(p)F_2(p) = \int_0^\infty \int_0^\infty e^{-p(\sigma + \tau)}f_1(\sigma)f_2(\tau)\,d\sigma\,d\tau.$$

Introducing the change of variable $\sigma = t - \tau$ and invoking the requirement that f_1 must vanish for negative values of its argument, we obtain

$$F_1(p)F_2(p) = \int_0^\infty e^{-pt}\left[\int_0^t f_1(t - \tau)f_2(\tau)\,d\tau\right]dt,$$

whence the transform of the quantity in brackets is $F_1 F_2$, the inversion of which yields (2.4.1).

We remark that, in typical applications, the right-hand side of (2.4.1) represents a superposition of effects of magnitude $f_2(\tau)$, arising at $t = \tau$, for which $f_1(t - \tau)$ is the *influence function*, i.e., the response to a unit impulse. Indeed, it constitutes the extension to impulsive inputs of Duhamel's superposition theorem for step inputs.

The unit impulse is known as the Dirac delta function and has the formal properties

$$\delta(t - \tau) = 0 \qquad (t \neq \tau) \tag{2.4.2a}$$

and

$$\int_{-\infty}^{\infty} f(\tau)\delta(t - \tau)\, d\tau = f(t). \tag{2.4.2b}$$

Setting $f(t) = \delta(t)$ in (2.2.1) and extending the integration to $t = 0-$, we obtain

$$\mathcal{L}\{\delta(t)\} = 1. \tag{2.4.3}$$

The delta function, as defined by (2.4.2), is improper, but it can be defined as the limit of a proper function and was so introduced by both Cauchy and Poisson. The function used by Cauchy was

$$\delta(x - a) = \frac{1}{\pi} \lim_{y \to 0+} \frac{y}{(x - a)^2 + y^2}, \tag{2.4.4}$$

the right-hand side of which may be identified as a solution to Laplace's equation for a doublet source located at $(x, y) = (a, 0)$. Perhaps the most satisfactory way of incorporating such functions as $\delta(t)$ in the application of integral transforms is provided by the theory of generalized functions [Erdélyi (1962) and Lighthill (1959)], but we rest content with the intuitive support provided by such physical idealizations as that of a concentrated force (which may be regarded as the limit of a large pressure distributed over a small area).

We illustrate the convolution theorem by replacing the constant force w_0 in (2.3.1) by an arbitrary force $w(t)$, and setting $x_0 = v_0 = 0$. Replacing the transform of w_0, w_0/p in (2.3.5), by $W(p)$, we obtain

$$X(p) = m^{-1}(p^2 + \beta^2)^{-1}W(p). \tag{2.4.5}$$

Setting $F_1 = m^{-1}(p^2 + \beta^2)^{-1}$ and $F_2 = W$ in (2.4.1) and inverting F_1 through T2.1.4, we obtain

$$x(t) = (m\beta)^{-1} \int_0^t w(\tau) \sin\left[\beta(t - \tau)\right] d\tau. \tag{2.4.6}$$

Setting $w = w_0$ (as a particular example) in (2.4.6), we recover the third term on the right-hand side of (2.3.8).

2.5 Heaviside's shifting theorem

Disturbances often arise at times other than zero (we may choose $t = 0$ to correspond to any particular event and include the effects of prior events in the initial conditions). Let $f(t - a)$ be a function that vanishes identically for $t < a$, where $a > 0$; then,

$$\mathcal{L}\{f(t - a)\} = \int_a^\infty e^{-pt}f(t - a)\, dt \qquad (2.5.1a)$$

$$= e^{-ap}\int_0^\infty e^{-pt}f(\tau)\, d\tau \qquad (2.5.1b)$$

$$= e^{-ap}F(p), \qquad (2.5.1c)$$

where (2.5.1b) follows from (2.5.1a) after the change of variable $\tau = t - a$, and (2.5.1c) follows from (2.5.1b) by virtue of (2.2.1), in which we may write τ as the variable of integration in place of t. Inverting (2.5.1c), we obtain

$$\mathcal{L}^{-1}\{e^{-ap}F(p)\} = \begin{cases} f(t - a) \\ 0 \end{cases} \qquad (t \gtrless a,\, a > 0). \qquad (2.5.2)$$

This is *Heaviside's shifting theorem.*

It often proves convenient, in dealing with discontinuities that arise at other than $t = 0$, to introduce the *Heaviside step function,*

$$H(t) = \begin{cases} 1 \\ 0 \end{cases} \qquad (t \gtrless 0). \qquad (2.5.3)$$

This introduction permits a result like (2.5.2) to be exhibited in the more compact form

$$\mathcal{L}^{-1}\{e^{-ap}F(p)\} = f(t - a)H(t - a). \qquad (2.5.4)$$

A related theorem, which is complementary to (2.5.1), is

$$\mathcal{L}^{-1}\{F(p + \alpha)\} = e^{-\alpha t}f(t). \qquad (2.5.5)$$

The proof follows directly from the inversion integral

$$f(t) = (2\pi i)^{-1}\int_{c - i\infty}^{c + i\infty} F(p)e^{pt}\, dp, \qquad (2.5.6)$$

together with an appropriate change of variable (the student should fill in the details). The parameter α may be complex provided that the parameter c in the corresponding inversion integral is appropriately chosen.

Suppose, for example, that the constant force in the problem of Section 2.3 is removed at $t = T$. Then,

$$w(t) = w_0[H(t) - H(t - T)] \qquad (2.5.7)$$

describes the applied force as a function of time. Replacing w_0/p in (2.3.3) and (2.3.5) by

$$W(p) = w_0 p^{-1}(1 - e^{-pT}), \qquad (2.5.8)$$

we obtain

$$x(t) = \frac{w_0}{k}\{(1 - \cos \beta t)H(t) - [1 - \cos \beta(t - T)]H(t - T)\} \quad (2.5.9)$$

as the response of the oscillator to the applied force. We emphasize that this example is intended only to illustrate the mechanics of the shifting theorem, since we could have obtained (2.5.9) directly from the super-position theorem for the oscillator. We use the shifting theorem to greater advantage in the following section.

2.6 Periodic functions

Let $f(t)$ be a periodic function of period T for $t > 0$, such that

$$f(t + T) = f(t) \quad (t > 0) \quad (2.6.1)$$

and $f(t)$ is piecewise continuous in $0 \leq t \leq T$. We then may calculate $F(p)$ as follows:

$$F(p) = \int_0^\infty e^{-pt}f(t)\,dt = \sum_{n=0}^\infty \int_{nT}^{(n+1)T} e^{-pt}f(t)\,dt \quad (2.6.2a)$$

$$= \sum_{n=0}^\infty e^{-npT} \int_0^T e^{-p\tau}f(\tau)\,d\tau \quad (2.6.2b)$$

$$= (1 - e^{-pT})^{-1} \int_0^T e^{-pt}f(t)\,dt, \quad (2.6.2c)$$

where (2.6.2b) follows from (2.6.2a) after the change of variable $t = nT + \tau$, and (2.6.2c) follows from the known result for the sum of a geometric series and the change of variable $\tau = t$.

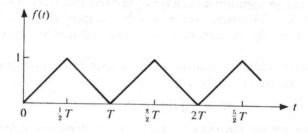

FIGURE 2.2   The triangular wave of Section 2.6.

Consider, for example, the triangular wave sketched in Figure 2.2 and described either by

$$f(t) = \begin{cases} \dfrac{2t}{T} & (0 \leq t \leq \tfrac{1}{2}T) \\[2ex] 2\left(1 - \dfrac{t}{T}\right) & (\tfrac{1}{2}T \leq t \leq T) \end{cases} \qquad (2.6.3)$$

in conjunction with (2.6.1) or by

$$f(t) = \frac{2}{T}\left[tH(t) + 2 \sum_{n=1}^{\infty} (-)^n (t - \tfrac{1}{2}nT)H(t - \tfrac{1}{2}nT) \right]. \qquad (2.6.4)$$

Substituting (2.6.3) into (2.6.2c), carrying out the integration, and cancelling the common factor $1 - \exp(-\tfrac{1}{2}pT)$ in the numerator and denominator of the result, we obtain

$$F(p) = \frac{2}{p^2 T} \tanh(\tfrac{1}{4}pT). \qquad (2.6.5)$$

Transforming (2.6.4) with the aid of T2.1.1 and the shifting theorem, T2.2.5, we obtain

$$F(p) = \frac{2}{p^2 T}\left[1 + 2 \sum_{n=1}^{\infty} (-)^n \exp(-\tfrac{1}{2}npT) \right] \equiv \frac{2}{p^2 T} S(p), \qquad (2.6.6)$$

where $S(p)$ denotes the bracketed series. The student may establish the equivalence of (2.6.5) and (2.6.6) by expanding the hyperbolic tangent in (2.6.5) in powers of $\exp(-\tfrac{1}{2}npT)$.

The inverse transform of (2.6.6) obviously yields (2.6.4). An alternative representation is provided by the expansion [Jolley (1961)]

$$\tanh y = 2y \sum_{s=1,3,\dots} [(\tfrac{1}{2}s\pi)^2 + y^2]^{-1}, \qquad (2.6.7)$$

the substitution of which into (2.6.5) yields

$$F(p) = \frac{16}{T^2} \sum_s p^{-1}(p^2 + \beta_s^2)^{-1}, \qquad (2.6.8a)$$

where

$$\beta_s = \frac{2\pi s}{T}, \qquad (2.6.8b)$$

and, here and throughout this section, s is summed over $1, 3, 5, \dots, \infty$. Remarking that each of the individual terms in (2.6.8a) is similar to the transform of (2.3.6), we obtain the inverse transform

$$f(t) = \frac{16}{T^2} \sum_s \beta_s^{-2}(1 - \cos \beta_s t) \qquad (2.6.9a)$$

$$= \frac{1}{2} - \frac{4}{\pi^2} \sum_s \frac{1}{s^2} \cos \frac{2s\pi t}{T}, \qquad (2.6.9b)$$

where (2.6.9b) follows from (2.6.9a) after invoking (2.6.8b) and the known result

$$\sum_s \frac{1}{s^2} = \frac{\pi^2}{8}$$

[which may be derived by letting $y \to 0$ in (2.6.7)]. We may identify (2.6.9b) as the Fourier-series representation of the periodic function $f(t)$, as defined by (2.6.1) and (2.6.3).

The response, say $x(t)$, of a linear system to a periodic input, say $f(t)$, may be determined by either of two, more or less distinct, procedures: (i) expanding $f(t)$ in a Fourier series, determining $x(t)$ for a particular Fourier component of $f(t)$, and summing over all components (which is permissible by virtue of linearity); (ii) determining $x(t)$ for $nT < t < (n + 1)T$, say $x_n(t)$, and then determining the constants of integration in $x_n(t)$ by invoking the appropriate continuity requirements. The Laplace-transform determination of $x(t)$ effectively combines these two procedures. Let us suppose, for example, that the constant force in the example of Section 2.3 is replaced by the periodic force

$$w(t) = w_1 f(t), \qquad (2.6.10)$$

where $f(t)$ is given by (2.6.3), and w_1 is the peak force ($w = w_1$ at $t = \frac{1}{2}T$). Replacing w_0/p by $w_1 F(p)$, setting $x_0 = v_0 = 0$ in (2.3.5), and introducing

$$x_1 = \frac{w_1}{k} \equiv \frac{w_1}{m\beta^2}, \qquad (2.6.11)$$

we obtain

$$X(p) = \beta^2 x_1 (p^2 + \beta^2)^{-1} F(p). \qquad (2.6.12)$$

The parameter x_1 is the static displacement of the oscillator under the action of the peak force w_1.

Considering first the representation (2.6.6) for $F(p)$ in (2.6.12), we obtain

$$X(p) = \frac{2\beta^2 x_1}{T} p^{-2}(p^2 + \beta^2)^{-1} S(p) \qquad (2.6.13a)$$

$$= \frac{2x_1}{T} [p^{-2} - (p^2 + \beta^2)^{-1}] S(p). \qquad (2.6.13b)$$

Inverting (2.6.13b) through T2.1.1, T2.1.4, and the shifting theorem, T2.2.5, we obtain

$$x(t) = \xi(t)H(t) + 2 \sum_{n=1}^{\infty} (-)^n \xi(t - \tfrac{1}{2}nT)H(t - \tfrac{1}{2}nT), \quad (2.6.14)$$

where

$$\xi(t) = \frac{2x_1}{T}(t - \beta^{-1} \sin \beta t) \quad (2.6.15)$$

is the inverse transform of (2.6.13b) for $S = 1$, corresponding to the leading term in the bracketed series of (2.6.6). We remark that the sum of the terms in t and $t - \tfrac{1}{2}nT$ in (2.6.14) is equal to $w_1 f(t)/k$, the static displacement of the oscillator under the force $w(t)$.

The representation (2.6.14) is especially advantageous for relatively small values of t, but it is not satisfactory for $t \gg T$ in consequence of the large number of terms that must be retained in the series;† moreover, (2.6.14) obscures the resonances that occur if $\beta T/2\pi$ is an odd integer. We therefore turn to the representation (2.6.8) for $F(p)$, the substitution of which into (2.6.12) yields

$$X(p) = \frac{16\beta^2 x_1}{T^2} \sum_s p^{-1}(p^2 + \beta_s^2)^{-1}(p^2 + \beta^2)^{-1} \quad (2.6.16a)$$

$$= \frac{16\beta^2 x_1}{T^2} \sum_s p^{-1}(\beta_s^2 - \beta^2)^{-1}[(p^2 + \beta^2)^{-1} - (p^2 + \beta_s^2)^{-1}] \quad (2.6.16b)$$

after a partial-fraction expansion. Inverting the individual terms in (2.6.16b) with the aid of (2.3.6), we obtain

$$x(t) = \frac{16\beta^2 x_1}{T^2} \sum_s (\beta_s^2 - \beta^2)^{-1}[\beta^{-2}(1 - \cos \beta t) - \beta_s^{-2}(1 - \cos \beta_s t)]. \quad (2.6.17)$$

Invoking the result

$$\frac{16}{T^2} \sum_s (\beta_s^2 - \beta^2)^{-1} = \sum_s [(\tfrac{1}{2}s\pi)^2 - (\tfrac{1}{4}\beta T)^2]^{-1} = 2(\beta T)^{-1} \tan(\tfrac{1}{4}\beta T), \quad (2.6.18)$$

which follows from (2.6.7) with $y = \tfrac{1}{4}i\beta T$ therein, we may place (2.6.17) in the alternative form

$$x(t) = \frac{2x_1}{\beta T} \tan(\tfrac{1}{4}\beta T)(1 - \cos \beta t) - \frac{16\beta^2 x_1}{T^2} \sum_s \beta_s^{-2}(\beta_s^2 - \beta^2)^{-1}(1 - \cos \beta_s t). \quad (2.6.19)$$

---

†It is, in fact, possible to sum the series in (2.6.14), but this reflects the simplicity of the particular example rather than any general property of such problems.

Invoking the result

$$\frac{16\beta^2}{T^2} \sum_s \beta_s^{-2}(\beta_s^2 - \beta^2)^{-1} = \frac{16}{T^2} \sum_s [(\beta_s^2 - \beta^2)^{-1} - \beta_s^{-2}]$$

$$= 2(\beta T)^{-1} \tan(\tfrac{1}{4}\beta T) - \frac{1}{2}, \qquad (2.6.20)$$

which follows from (2.6.18), and substituting β_s from (2.6.8b) into the remaining series, we obtain the additional representation

$$x(t) = x_1 \left\{ \frac{1}{2} - 2(\beta T)^{-1} \tan(\tfrac{1}{4}\beta T) \cos \beta t \right.$$

$$\left. + \left(\frac{2\beta T}{\pi}\right)^2 \sum_s s^{-2}[(2\pi s)^2 - (\beta T)^2]^{-1} \cos \left(\frac{2\pi s t}{T}\right) \right\}. \qquad (2.6.21)$$

The terms in $\cos \beta t$ in each of (2.6.17), (2.6.19), and (2.6.21) represent the free oscillation that must be added to the forced oscillation to satisfy the initial conditions; the remaining terms represent the forced oscillation produced by the periodic force. Resonance occurs if $\beta T/2\pi$ is an odd integer, say s_*, in which case both $\tan(\tfrac{1}{4}\beta T)$ and the s_*th term in each of the series are infinite. Applying L'Hospital's rule to the s_*th term in (2.6.17), we obtain

$$x_*(t) = \frac{16 x_1}{\beta^2 T^2} [1 - \cos \beta t - \tfrac{1}{2}\beta t \sin \beta t] \qquad (\beta T = 2\pi s_*). \quad (2.6.22)$$

We could have dealt with the problem of resonance directly in (2.6.16a), for which the s_*th term is

$$X_* = \frac{16 \beta^2 x_1}{T^2} p^{-1}(p^2 + \beta^2)^{-2} \qquad (\beta T = 2\pi s_*). \qquad (2.6.23)$$

Differentiating (2.3.6) with respect to the parameter β, we find that the inverse transform of (2.6.23) is given by (2.6.22).

2.7 The inversion integral

The most expedient procedure for the inversion of a given transform is through a suitable table of transform pairs, such as (but typically much more extensive than) that of Table 2.1, augmented by operational theorems, such as those of Table 2.2. This procedure is illustrated by the examples in the preceding sections and in several of the exercises at the end of this chapter. It is clearly limited by the extent of the available tables, although it typically suffices for the solution of ordinary differential equations with constant coefficients, such as those that arise in electric-circuit and mechanical-vibration problems.

The most general procedure is through the inversion integral,

$$f(t) = \mathcal{L}^{-1}\{F(p)\} \equiv \frac{1}{2\pi i} \int_{c-i\infty}^{c+i\infty} F(p)e^{pt}\, dp, \qquad (2.7.1)$$

in which $F(p)$ is a prescribed function of the complex variable p that has no singularities in $c \leq p_r < \infty$. We assume that $t > 0$; the path of integration could be closed in $p_r > c$ if $t < 0$, with the result that $f(t) = 0$ by virtue of the fact that $F(p)$ would have no singularities within the resulting, closed contour.

We consider first that class of problems in which $F(p)$ is a *meromorphic function*—that is, a function having no singularities other than poles for $|p| < \infty$—and is bounded at infinity according to

$$\lim_{|p| \to \infty} |p|^b F(p) < \infty \qquad (b > 0,\ p \neq p_n), \qquad (2.7.2)$$

where $p_1, p_2, \ldots$ are the poles of $F(p)$. Poles rarely occur in $p_r > 0$, although they frequently occur on $p_r = 0$; accordingly, it usually suffices to let $c \to 0+$. Typical Laplace transforms (of the type considered herein) decay at least like $1/|p|$ as $|p| \to \infty$, corresponding to $b \geq 1$ in (2.7.2), but it is useful to have the greater generality implied by the weaker restriction $b > 0$.

Consider the contour integral

$$I = \frac{1}{2\pi i} \int_C F(p)e^{pt}\, dp, \qquad (2.7.3)$$

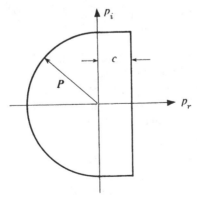

FIGURE 2.3   Contour of integration for meromorphic Laplace transform.

where (see Figure 2.3) C consists of a vertical segment, $p_r = c$, $-P \leqq p_i \leqq P$, and a semicircular arc of radius P in $p_r < c$ that does not intersect any of the poles of $F(p)$. The integral over this arc vanishes in the limit $P \to \infty$ provided that $F(p)$ satisfies the conditions assumed above—in particular (2.7.2); the proof of this statement, which constitutes an extension of Jordan's lemma [which would be directly applicable for $c = 0+$ in (2.7.1) and $b \geqq 1$ in (2.7.2)] is given by Carslaw–Jaeger (1953, Section 31). The integral over the vertical segment in this same limit tends to $f(t)$, as defined by (2.7.1); accordingly, $I \to f(t)$ as $P \to \infty$. It then follows from Cauchy's residue theorem that $f(t)$ is given by $2\pi i$ times the sum of the residues of $F(p)e^{pt}$ at its poles, say $p_1, p_2, \ldots$, which, by hypothesis, lie in $p_r < c$. If, as is typically true, these poles are all simple, the result is

$$f(t) = \sum_n R(p_n) \exp(p_n t), \qquad (2.7.4a)$$

where

$$R(p_n) = \lim_{p \to p_n} (p - p_n)F(p), \qquad (2.7.4b)$$

the summation is over all of the poles, and $R(p_n)$ is the residue of $F(p)$ at $p = p_n$.

It frequently proves convenient to place $F(p)$ in the form

$$F(p) = \frac{G(p)}{H(p)}, \qquad (2.7.5)$$

where $G(p)$ has no singularities in $|p| < \infty$ [but see sentence following (2.7.6)], and $H(p)$ has simple zeros at $p = p_1, p_2, \ldots$. Substituting (2.7.5) into (2.7.4b) and invoking L'Hospital's rule, we obtain

$$R(p_n) = \lim_{p \to p_n} \frac{(p - p_n)G(p)}{H(p)} = \frac{G(p_n)}{H'(p_n)}, \qquad (2.7.6)$$

where $H(p_n) \equiv 0$, and $H'(p_n)$ is the derivative of $H(p)$ at $p = p_n$. The result (2.7.6) may be generalized to allow any convenient factoring of the form (2.7.5) in the neighborhood of a given pole, $p = p_n$, subject only to the requirement that $G(p)$ be regular in that neighborhood. Substituting (2.7.5) and (2.7.6) into (2.7.4a), we obtain

$$\mathscr{L}^{-1} \frac{G(p)}{H(p)} = \sum_n \frac{G(p_n)}{H'(p_n)} \exp(p_n t), \qquad \text{where} \qquad H(p_n) = 0. \quad (2.7.7)$$

This last result is due essentially to Heaviside (who gave an equivalent form) and is generally known as the *Heaviside expansion theorem*.

The zeros of $H(p)$ are typically either real or complex-conjugate pairs. The net contribution of such a pair to the summation of either (2.7.4a) or (2.7.7) is equal to twice the real part of either complex zero taken separately.

It occasionally happens that $H(p)$ has a double zero. The required modification of (2.7.7) is given in Exercise 2.8. Zeros of third or higher order rarely occur at other than $p = 0$ and are best handled on an ad hoc basis, rather than by further generalizations of (2.7.7).

We illustrate (2.7.7) by returning to the transform given by (2.6.5) and (2.6.12),

$$X(p) = \frac{2\beta^2 x_1}{T} p^{-2}(p^2 + \beta^2)^{-1} \tanh(\tfrac{1}{4}pT), \qquad (2.7.8)$$

which is a meromorphic function that has simple poles at

$$p = 0, \ \pm i\beta, \ \pm \frac{2\pi i s}{T} \equiv \pm i\beta_s \qquad (s = 1, 3, 5, \ldots, \infty) \qquad (2.7.9)$$

and vanishes like $|p|^{-4}$ as $|p| \to \infty$; accordingly, it satisfies the required conditions for the validity of (2.7.7).† We may place it in the form of (2.7.5) by choosing

$$G(p) = \frac{2\beta^2 x_1}{pT} \sinh(\tfrac{1}{4}pT), \qquad H(p) = p(p^2 + \beta^2) \cosh(\tfrac{1}{4}pT). \quad (2.7.10)$$

We emphasize that G is regular at $p = 0$. Substituting (2.7.10) into (2.7.6), we obtain

$$R(0) = \tfrac{1}{2}x_1, \qquad R(\pm i\beta) = -x_1(\beta T)^{-1} \tan(\tfrac{1}{4}\beta T),$$

and

$$R(\pm i\beta_s) = \frac{8\beta^2 x_1}{T^2} \beta_s^{-2}(\beta_s^2 - \beta^2)^{-1}. \qquad (2.7.11)$$

Substituting (2.7.11) into (2.7.4a) and collecting the contributions of the complex poles in conjugate pairs, we obtain (2.6.21).

We turn now to that more general class of problems in which $F(p)$ may have not only poles, but also branch points, in $p_r < c.$‡ The contour C for the integral of (2.7.3) then must be deformed around the appropriate

---

†If the derivation of (2.7.4a) had been given explicitly for $X(p)$, it would have been expedient to have assumed $P = (N + \frac{1}{2})(\pi/T)$, where N is an integer, and let $P \to \infty$ by letting $N \to \infty$ in order to avoid the difficulty of having C pass through a pole.

‡The remaining material in this section is not required in Section 2.8, and the student may find it expedient to read Sections 2.8 and 2.9 through (2.9.7) before studying this material in detail.

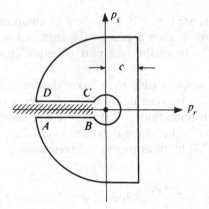

FIGURE 2.4 Contour of integration for Laplace transform having branch
point at the origin.

branch cuts, as illustrated in Fig. 2.4 for the important special case of a
branch point at $p = 0$. More generally, branch points of Laplace transforms are likely to be on the imaginary axis, but only occasionally else-
where. The contributions of the poles, if any, may be evaluated as before,
in particular from (2.7.4), but it also is necessary to include the integral
around the branch cut.

Considering the contour of Figure 2.4, we let $p = ue^{-i\pi}$ on $AB, p = \varepsilon e^{i\theta}$
on BC, and $p = ue^{i\pi}$ on CD to obtain the contributions

$$I_{AB} = \frac{1}{2\pi i} \int_{\infty}^{\varepsilon} e^{-ut}F(ue^{-i\pi})(-du),$$

$$I_{BC} = \frac{1}{2\pi i} \int_{-\pi}^{\pi} \exp(\varepsilon t e^{i\theta})F(\varepsilon e^{i\theta})(i\varepsilon e^{i\theta}\, d\theta),$$

and

$$I_{CD} = \frac{1}{2\pi i} \int_{\varepsilon}^{\infty} e^{-ut}F(ue^{i\pi})(-du)$$

to the contour integral I [note that $p = -u$ on both sides of the cut,
whereas $F(ue^{i\pi})$ and $F(ue^{-i\pi})$ have different values if, as assumed, $p = 0$
is a branch point]. The integral I_{BC} vanishes in the limit $\varepsilon \to 0$ if $pF(p) \to 0$
as $p \to 0$; otherwise it must be evaluated by expanding the integrand
about $\varepsilon = 0$ on a more or less ad hoc basis [note that $e^{pt}F(p)$ does *not*
have a Laurent-series representation in the neighborhood of $p = 0$]. If,
as in the example of Section 2.9 below, $F(p)$ is of the form $p^{-1}F_1(p)$, where

$F_1(p)$ has a branch point at $p = 0$ but is finite there, the limit of I_{BC} as $\varepsilon \to 0$ is simply $F_1(0)$. If $F(p)$ has an infinity of higher order, the limiting forms of each of I_{AB}, I_{BC}, and I_{CD} may be improper, in which case the limit as $\varepsilon \to 0$ may be taken only after the three integrals have been combined.

Proceeding on the assumption that I_{BC} exists, we combine I_{AB} and I_{CD} and write

$$J \equiv \lim_{\varepsilon \to 0} (I_{AB} + I_{CD}) = \int_0^\infty e^{-ut}\phi(u) \, du, \qquad (2.7.12a)$$

wherein

$$\phi(u) = (2\pi i)^{-1}[F(ue^{-i\pi}) - F(ue^{i\pi})]. \qquad (2.7.12b)$$

It may be possible to regard $J(t)$ as the Laplace transform of $\phi(u)$, with t as the transform variable, and to evaluate it from a table of Laplace transforms; however, whether or not this is true, it often suffices to obtain an asymptotic approximation for large t (such approximations sometimes prove satisfactory for surprisingly small values of t).

The asymptotic evaluation of J is especially simple if $\phi(u)$ has the representation (in $u < R$)

$$\phi(u) = \sum_{n=1}^\infty a_n u^{(n/r)-1} \qquad (r > 0, u < R), \qquad (2.7.13a)$$

wherein $r = 1$ or 2 in typical applications. If positive real numbers C and a exist such that $|\phi| < Ce^{au}$, $u > R$, then *Watson's lemma* states that J has the asymptotic expansion

$$J \sim \sum_{n=1}^\infty a_n \Gamma\left(\frac{n}{r}\right) t^{-n/r} \qquad (t \to \infty), \qquad (2.7.13b)$$

where Γ denotes the gamma function. See Copson (1965, Section 22) for the proof.

Consider, for example, $F(p) = p^{1/2}$; then

$$\phi(u) = (2\pi i)^{-1}[u^{1/2}e^{-i\pi/2} - u^{1/2}e^{i\pi/2}] = -\pi^{-1}u^{1/2}.$$

Setting $r = 2$, $n = 3$, and $a_2 = -1/\pi$ in (2.7.13), we obtain

$$J \sim -\pi^{-1}\Gamma(\tfrac{3}{2})t^{-3/2} = -\tfrac{1}{2}(\pi t^3)^{-1/2}.$$

More generally (see Exercise 2.19),

$$F(p) = \sum_{n=1}^\infty b_n p^{(n/r)-1} \qquad (r > 0) \qquad (2.7.14a)$$

implies

$$J \sim \pi^{-1} \sum_{n=1}^{\infty} b_n \Gamma\left(\frac{n}{r}\right) \sin \frac{\pi n}{r} t^{-n/r} \qquad (t \to \infty), \qquad (2.7.14b)$$

to which terms for which n/r is an integer make no contribution.

We obtain the complementary result that

$$F(p) \sim \sum_{n=1}^{\infty} c_n p^{-n/r} \qquad (|p| \to \infty, \ r > 0) \qquad (2.7.15a)$$

implies

$$f(t) = \sum_{n=1}^{\infty} \frac{c_n}{\Gamma(n/r)} t^{(n/r)-1} \qquad (t \to 0) \qquad (2.7.15b)$$

by applying Watson's lemma directly to the Laplace-transform integral.

The expansion of $\phi(x)$ about $x = 0$ may contain logarithmic terms, in which case a formal asymptotic expansion may be obtained by integrating term by term and invoking the result (see Exercise 2.13)

$$\int_0^{\infty} e^{-xt} x^m \log x \, dx = m!(1 + \tfrac{1}{2} + \tfrac{1}{3} + \cdots + m^{-1} - \gamma - \log t)t^{-m-1},$$

$$(2.7.16)$$

where $\gamma = 0.577215\ldots$ is Euler's constant and $1 + \cdots + m^{-1} \equiv 0$ for $m = 0$.

Other methods of evaluating branch-cut integrals, of which J is only one—albeit the most important—form, are discussed in the monographs of Erdélyi (1956) and Copson (1965). Finally, we note that the numerical evaluation of integrals such as J may be entirely practical by virtue of the exponential convergence, although it may be necessary first to separate the contribution of the singularity at the origin.

2.8 Wave propagation in a bar †

A uniform bar of cross section A is at rest and unstressed for $t < 0$, with one end fixed at $x = 0$ (see Figure 2.5). At $t = 0$, a force of magnitude PA is applied to the free end, $x = l$, in the direction of the positive x axis. We require the subsequent motion on the hypothesis of small displacements.

Let $y(x, t)$ denote the longitudinal displacement from equilibrium of the section at x. Hooke's law implies that the stress associated with the strain y_x at any section is Ey_x, so that the net force on a differential element of length dx and area A between x and $x + dx$ (see Figure 2.6) is

---

† Carslaw–Jaeger (1953), Morse (1948), and Sneddon (1951) give many applications of the Laplace transform to problems of mechanical wave propagation.

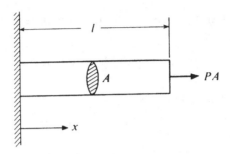

FIGURE 2.5 A uniform bar to which a load PA is applied abruptly at time $t = 0$.

$$AE[y_x + y_{xx}\, dx] - AEy_x = AEy_{xx}\, dx,$$

where E is Young's modulus, and subscripts denote partial differentiation. The mass of that element is $\rho A\, dx$, where ρ is its density. Equating the product of this mass and the acceleration to the net force on the section and dividing the result by $\rho A\, dx$, we obtain the wave equation

$$c^2 y_{xx} = y_{tt}, \qquad \text{where} \qquad c = \left(\frac{E}{\rho}\right)^{1/2}. \tag{2.8.1}$$

The initial conditions are

$$y = y_t = 0 \qquad \text{at } t = 0 \text{ and } 0 < x < l. \tag{2.8.2}$$

The boundary conditions are

$$y = 0 \qquad \text{at } x = 0 \tag{2.8.3a}$$

and

$$Ey_x = P \qquad \text{at } x = l \qquad (t > 0). \tag{2.8.3b}$$

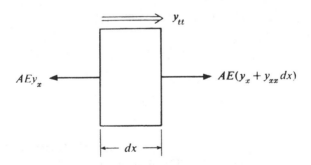

FIGURE 2.6 An element of the bar, showing the forces on the faces at x and $x + dx$ and the acceleration vector, y_{tt}.

Taking the Laplace transform of (2.8.1) with respect to t and invoking (2.8.2), we obtain

$$Y_{xx} - \left(\frac{p}{c}\right)^2 Y = 0. \tag{2.8.4}$$

Transforming (2.8.3), we obtain

$$Y = 0 \qquad \text{at } x = 0 \tag{2.8.5a}$$

and

$$Y_x = \frac{P}{Ep} \qquad \text{at } x = l. \tag{2.8.5b}$$

The most general solution of the ordinary differential equation (2.8.4) is

$$A \sinh \frac{px}{c} + B \cosh \frac{px}{c}.$$

We choose $B = 0$ in order to satisfy (2.8.5a) and then determine A to satisfy (2.8.5b). The result is

$$Y = \frac{Pc \sinh(px/c)}{Ep^2 \cosh(pl/c)}, \tag{2.8.6}$$

as may be verified by direct substitution. Remarking that $Y(p)$ has simple poles at

$$p = 0, \qquad \frac{\pm(2n+1)i\pi c}{2l} \qquad (n = 0, 1, 2, \ldots), \tag{2.8.7}$$

we may invert (2.8.6) through (2.7.7) by choosing

$$G(p) = \frac{Pc}{Ep} \sinh \frac{px}{c}, \qquad H(p) = p \cosh \frac{pl}{c}. \tag{2.8.8}$$

The end result is (the student should fill in the details)

$$y(x, t) = \frac{Pl}{E}\left[\frac{x}{l} - \frac{8}{\pi^2} \sum_{n=0}^{\infty} (-)^n(2n+1)^{-2} \sin(k_n x) \cos(k_n ct)\right], \tag{2.8.9}$$

where

$$k_n = \frac{(2n+1)\pi}{2l}. \tag{2.8.10}$$

The first term in (2.8.9), Px/E, represents the ultimate (static) displacement of the bar; the remaining terms represent standing waves that would die out gradually if friction were admitted. In the absence of friction, however, the displacement continues to oscillate about the static displacement.

We obtain an alternative solution by expressing the hyperbolic functions in terms of exponentials and substituting the expansion

$$\frac{1}{\cosh(pl/c)} = \frac{2\exp(-pl/c)}{1 + \exp(-2pl/c)} = 2\sum_{n=0}^{\infty} (-)^n \exp\left[-\frac{(2n+1)pl}{c}\right] \quad (2.8.11)$$

into (2.8.6) to obtain

$$Y = \frac{Pc}{Ep^2} \sum_{n=0}^{\infty} (-)^n$$

$$\times \left\{ \exp\left[-\frac{p}{c}((2n+1)l - x)\right] - \exp\left[-\frac{p}{c}((2n+1)l + x)\right] \right\}. \quad (2.8.12)$$

Now the inverse transform of p^{-2} is t, whence the shifting theorem, T2.2.5, yields

$$y(x,t) = \frac{P}{E} \sum_{n=0}^{\infty} (-)^n \{[ct - ((2n+1)l - x)] - [ct - ((2n+1)l + x)]\}, \quad (2.8.13)$$

where, by definition, the square brackets vanish identically if their contents are negative.

Equation (2.8.13) exhibits the solution as a series of traveling waves, the first and second sets moving respectively toward and away from $x = 0$. Such a representation is valuable not only because it presents the solution in a finite number of terms (since each of only a finite number of the square brackets is positive at any finite time), thereby rendering numerical computation simpler for small ct/l, but also because it provides additional insight into the physical problem. It is, indeed, one of the great virtues of the Laplace-transform solution that it comprises both the standing- and traveling-wave representations (cf. Section 2.6).

At $x = l$, where the load acts, the displacement given by (2.8.13) reduces to

$$y(l,t) = \frac{Pc}{E}\left\{ [t] + 2\sum_{n=1}^{\infty} (-)^n \left[t - \frac{2nl}{c}\right] \right\}, \quad (2.8.14)$$

which is simply the triangular wave of Figure 2.2, with a mean value equal to the static displacement, Pl/E (so that the maximum value is $2Pl/E$, rather than unity as in Figure 2.2), and a period equal to the time required for a wave to travel four times the length of the bar, $T = 4l/c$.

2.9 Heat conduction in a semi-infinite solid†

We consider now the classical problem of a semi-infinite solid, $x > 0$, that is initially at temperature $v = 0$ and for which the boundary, $x = 0$, is maintained at temperature $v = v_0$ for $t > 0$. The rate at which heat is transferred across a plane section, $x = $ const, in the direction of increasing x is $-Kv_x$, where K is the thermal conductivity; accordingly the net rate at which heat is transferred *into* a slab bounded by x and $x + dx$ is $Kv_{xx}\, dx$ per unit area. This must be equal to the rate at which the slab is gaining heat, namely $\rho c v_t\, dx$, where ρ is the density, c is the specific heat, and dx is the volume (per unit of transverse area). Equating these two rates, we obtain the diffusion equation

$$v_t = \kappa v_{xx}, \qquad \text{where} \qquad \kappa = \frac{K}{\rho c} \qquad (2.9.1)‡$$

is the *diffusivity*. The initial condition is

$$v = 0 \qquad \text{at } t = 0 \quad (x > 0). \qquad (2.9.2)$$

The boundary conditions are

$$v = v_0 \qquad \text{at } x = 0 \qquad (2.9.3a)$$

and

$$|v| < \infty \qquad \text{as } x \to \infty. \qquad (2.9.3b)$$

Transforming (2.9.1)–(2.9.3), we obtain

$$V_{xx} - \frac{p}{\kappa} V = 0, \qquad (2.9.4)$$

$$V = \frac{v_0}{p} \qquad \text{at } x = 0, \qquad (2.9.5a)$$

and

$$|V| < \infty \qquad \text{as } x \to \infty. \qquad (2.9.5b)$$

Choosing that exponential solution of (2.9.4) which satisfies (2.9.5b) and then choosing its amplitude to satisfy (2.9.5a), we obtain

$$V = \frac{v_0}{p} \exp\left[-\left(\frac{p}{\kappa}\right)^{1/2} x \right], \qquad (2.9.6)$$

where $(p^{1/2})_r > 0$ for $p_r > 0$. Invoking T2.1.10, we obtain

---

† Carslaw–Jaeger (1949) give extensive applications of the Laplace transform to heat-conduction problems in one, two, or three dimensions.

‡ The right-hand side of (2.9.1) becomes $\kappa \nabla^2 v$ for three-dimensional (isotropic) heat conduction.

$$\frac{v}{v_0} = \text{erfc}[\tfrac{1}{2}(\kappa t)^{-1/2}x] \equiv 2\pi^{-1/2} \int_{(\kappa t)^{-1/2}(x/2)}^{\infty} e^{-u^2} \, du, \qquad (2.9.7)$$

where erfc denotes the complementary error function.

The error function, $\text{erf } z \equiv 1 - \text{erfc } z$, is tabulated, and its expansions for both small and large z are known; however, we use the transform of (2.9.6) to illustrate the general procedure for multivalued transforms described in Section 2.7. We choose the branch cut for $p^{1/2}$ along the negative-real axis to satisfy $(p^{1/2})_r > 0$ [the student should verify this by setting $p = ue^{i\theta}$ and showing $(p^{1/2})_r = u^{1/2}\cos(\theta/2) > 0$ in $|\theta| < \pi$] and close the contour of integration in $p_r < 0$, as shown in Figure 2.4. The function $V(p)$ has no singularities inside the closed contour, whence the contour integral of $V(p) \exp(pt)$ around that contour is zero. Moreover, $F(p)$ satisfies (2.7.2) with $b = 1$ therein, whence the integrals over the arcs at infinity also are zero. It follows that the inverse transform of V is equal to the integral over the top and bottom of the branch cut plus the integral around the small circle at the origin taken in a *counterclockwise* direction,

$$v = \frac{v_0}{2\pi i} \int_{ABCD} p^{-1} \exp\left\{ pt - \left(\frac{p}{\kappa}\right)^{1/2} x \right\} dp, \qquad (2.9.8)$$

where the contour $ABCD$ is as in Figure 2.4.

In the neighborhood of the origin, V tends to infinity like v_0/p, and the contribution of the path BC as its radius tends to zero is simply v_0 [as may be proved by setting $p = \varepsilon \exp(i\theta)$ in the integrand, letting $\varepsilon \to 0$, and integrating between $-\pi$ and π; cf. I_{BC} in Section 2.7]. Letting $F(p) = V(p)/v_0$ in the development of Section 2.7, we obtain

$$v = v_0(1 + J), \qquad (2.9.9)$$

where J is given by (2.7.12a) with

$$\phi(u) = \frac{1}{2\pi i} \left\{ (-u)^{-1} \exp\left[-\left(\frac{u}{\kappa}\right)^{1/2} xe^{-i\pi/2} \right] \right.$$

$$\left. - (-u)^{-1} \exp\left[-\left(\frac{u}{\kappa}\right)^{1/2} xe^{i\pi/2} \right] \right\}$$

$$= -\frac{1}{\pi u} \sin\left[\left(\frac{u}{\kappa}\right)^{1/2} x \right] \qquad (2.9.10a)$$

$$= -\frac{1}{\pi} \sum_{m=0}^{\infty} \frac{(-)^m}{(2m+1)!} \kappa^{-m-(1/2)} x^{2m+1} u^{m-(1/2)}. \qquad (2.9.10b)$$

Invoking (2.7.13), with $r = 2$, and substituting the result into (2.9.9), we obtain

$$\frac{v}{v_0} \sim 1 - \frac{1}{\pi} \sum_{m=0}^{\infty} \frac{(-)^m \Gamma(m + \frac{1}{2}) x^{2m+1}}{(2m + 1)!(\kappa t)^{m+(1/2)}} \qquad (t \to \infty). \qquad (2.9.11)$$

Finally, anticipating the alternative solution of Section 3.5 below, we substitute (2.9.10a) into (2.7.12a) and (2.9.9) and introduce the change of variable $u = \kappa k^2$ to obtain the representation

$$\frac{v}{v_0} = 1 - \frac{2}{\pi} \int_0^{\infty} k^{-1} \exp(-k^2 \kappa t) \sin kx \, dk. \qquad (2.9.12)$$

2.10 Oscillating airfoil in supersonic flow †

We consider (see Figure 2.7) the disturbances produced by a thin (two-dimensional) airfoil that is traveling through a perfect (inviscid, non-heat-conducting) fluid with the supersonic speed U and that executes small, transverse oscillations. The disturbances, being small by hypothesis, are sound waves (the flow is isentropic). The velocity potential for these disturbances, defined such that the perturbation velocity is given by

$$\mathbf{v} = \nabla \phi, \qquad (2.10.1)$$

satisfies the two-dimensional wave equation in a fixed reference frame (with respect to which the fluid at infinity is at rest),

$$c^2(\phi_{XX} + \phi_{yy}) = \phi_{TT}, \qquad (2.10.2)$$

where X and y are Cartesian coordinates, and T is time. This reference frame is not convenient for the description of the motion of the airfoil,

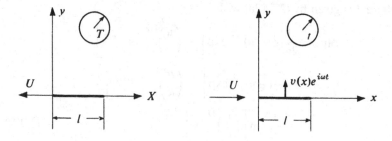

FIGURE 2.7   Schematic of oscillating airfoil in fixed (X, y, T) and moving reference frames.

---

†This section may be omitted without loss of continuity.

which is more conveniently described in a reference frame that moves with the mean velocity U; accordingly, we introduce the Galilean transformation

$$x = X + UT, \qquad t = T, \qquad (2.10.3)$$

under which (2.10.2) goes over to (the student should carry out this transformation in detail and observe that, whereas $t = T$, $\phi_t \neq \phi_T$; conversely, $x \neq X$, but $\phi_x = \phi_X$)

$$c^2(\phi_{xx} + \phi_{yy}) = \left(\frac{\partial}{\partial t} + U\frac{\partial}{\partial x}\right)^2 \phi. \qquad (2.10.4)$$

We assume that the transverse velocity of the airfoil is harmonic in time, with angular frequency ω, and has the complex amplitude $v(x)$, such that it is given by the real part of $v(x)\exp(i\omega t)$. We may evaluate the corresponding velocity in the air, ϕ_y, at $y = 0+$ $(0-)$ for points just above (below) the airfoil by virtue of the assumption that the airfoil is thin. We then may pose the boundary conditions for ϕ in the form

$$\phi_y = v(x)e^{i\omega t} \text{ on } y = 0\pm \text{ and } x > 0 \quad \text{and} \quad |\phi| < \infty \text{ as } |y| \to \infty, \qquad (2.10.5)$$

with the implicit understanding that the required solution is given by the real part of ϕ.

Invoking the fact that the airfoil is moving with supersonic speed, we infer that no disturbances can appear upstream of the airfoil, in consequence of which $\phi \equiv 0$ in $x < 0$. It then follows (since neither ϕ nor $\nabla\phi$ can change discontinuously) that

$$\phi = \phi_x = 0 \qquad \text{at } x = 0. \qquad (2.10.6)$$

These are, in effect, *initial conditions* with respect to the variable x, which now plays the same role as time in a conventional initial-value problem.

Invoking the linearity of (2.10.4) and (2.10.5) in ϕ, we assume that ϕ exhibits the time dependence $\exp(i\omega t)$, so that we may replace $(\partial/\partial t)$ by $i\omega$ in (2.10.4). Taking the Laplace transform of the resulting equation and of the boundary conditions (2.10.5) with respect to x and invoking the initial conditions (2.10.6), we obtain

$$\Phi_{yy} + p^2\Phi = \left(\frac{Up + i\omega}{c}\right)^2 \Phi \qquad (2.10.7)$$

and

$$\Phi_y = V(p)e^{i\omega t} \qquad \text{at } y = 0\pm. \qquad (2.10.8)$$

That solution of (2.10.7) which is bounded as $y \to \infty$ is proportional to $\exp(-\lambda y)$, where

$$\lambda = \left[\left(\frac{Up + i\omega}{c} \right)^2 - p^2 \right]^{1/2} \tag{2.10.9a}$$

$$\equiv B[(p + ivM)^2 + v^2]^{1/2}, \tag{2.10.9b}$$

$$M = \frac{U}{c}, \qquad B = (M^2 - 1)^{1/2}, \qquad v = \frac{\omega}{B^2 c}, \tag{2.10.10}$$

and that branch of the radical which renders $\lambda_r > 0$ for $p_r > 0$ is implied. Invoking (2.10.8), we obtain

$$\Phi = -\lambda^{-1} V(p) e^{i\omega t - \lambda y} \qquad (y > 0), \tag{2.10.11}$$

wherein the sign of λ must be reversed if $y < 0$.

We may invert Φ with the aid of the convolution theorem. Considering first the factor $\lambda^{-1} \exp(-\lambda y)$, we apply the shifting theorem to obtain

$$\mathscr{L}^{-1}\{\lambda^{-1} e^{-\lambda y}\} = B^{-1} e^{-ivMx} \mathscr{L}^{-1}\{(p^2 + v^2)^{-1/2} \exp[-(p^2 + v^2)^{1/2} By]\} \tag{2.10.12a}$$

$$= B^{-1} e^{-ivMx} J_0[v(x^2 - B^2 y^2)^{1/2}] H(x - By), \tag{2.10.12b}$$

where (2.10.12b) follows from (2.10.12a) through T2.1.12. Invoking the convolution theorem, we obtain

$$\phi = -B^{-1} H(x - By) \int_0^{x - By} K(x - \xi, y) v(\xi)\, d\xi, \tag{2.10.13}$$

where

$$K(x, y) = e^{i(\omega t - vMx)} J_0[v(x^2 - B^2 y^2)^{1/2}]. \tag{2.10.14}$$

This last example, which was originally solved by various workers with the aid of more difficult techniques [see Miles (1959), Section 5.2 for complete references and for other applications of transform methods to aerodynamic problems], illustrates the power of the Laplace transform when applied to problems of some complexity.

EXERCISES[†]

2.1 Derive T2.1.5 and T2.1.6 by: (a) considering α to be complex in T2.1.2 and taking the real and imaginary parts of the result; (b) applying the shifting theorem, T2.2.6, to T2.1.3 and T2.1.4.

[†]The starred exercises are of greater difficulty.

2.2 Derive T2.1.7 by: (a) differentiating T2.1.2 with respect to the parameter α; (b) applying the shifting theorem to T2.1.1.

2.3 Derive T2.1.1 for $v = n$ by differentiating T2.1.2 with respect to α.

2.4 Derive T2.1.2 by expanding the exponential in a power series, invoking T2.1.1, and summing the resulting series in p.

2.5 Derive T2.1.8 by: (a) transforming the derivative of the Heaviside step function, $H(t)$; (b) transforming the function

$$f = \begin{cases} 1/\varepsilon & (0 < t < \varepsilon) \\ 0 & (t > \varepsilon) \end{cases}$$

and considering the limit $\varepsilon \to 0+$.

2.6 Derive the scaling theorem, T2.2.7.

2.7 Derive T2.2.10 directly from the definition of $F(p)$.

2.8 Suppose that one of the zeros of $H(p)$ in (2.7.5), say p_1, is a double zero, such that $H(p_1) = H'(p_1) = 0$. Show that the first term in the series of (2.7.7) must be replaced by

$$[\Phi'(p_1) + \Phi(p_1)t] \exp(p_1 t),$$

where

$$\Phi(p) = \frac{(p - p_1)^2 G(p)}{H(p)}.$$

Use this result to invert (2.6.23).

2.9 Show that

$$\lim_{p \to 0+} pF(p) = \lim_{t \to \infty} f(t)$$

provided that both limits exist.

2.10 Use the convolution theorem, T2.2.9, to obtain a general solution of the second-order differential equation

$$f''(x) + af'(x) + bf(x) = g(x).$$

2.11 The rectified sine-wave voltage $v(t) = v_0|\sin \omega t|$ is applied to a series circuit consisting of a resistance R in series with an inductance L at $t = 0$. Use the methods of Section 2.6 to determine two different representations of the resulting current.

2.12 Suppose that a viscous damper is placed across the spring in Figure 2.1, such that the equation of motion becomes

$$mx'' + dx' + kx = w_0.$$

Show that the resulting motion is given by

$$x = e^{-\alpha t}[x_0 \cos \beta t + (\alpha x_0 + v_0)\beta^{-1} \sin \beta t] + \frac{w_0}{k}\left[1 - e^{-\alpha t}\left(\cos \beta t + \frac{\alpha}{\beta} \sin \beta t\right)\right]$$

where $\alpha = d/2m$ and $\beta^2 = (k/m) - \alpha^2 > 0$. Find the motion in the special case of critical damping, for which d is adjusted to give $\beta = 0$.

2.13   Use the known result

$$\gamma = -\int_0^\infty e^{-u} \log u \, du \qquad (\gamma \equiv 0.577\ldots)$$

to derive (2.7.16) for $m = 0$ and then generalize the result through integration by parts.

2.14   Derive T2.1.11 by choosing the branch cut for $F(p) = (p^2 + a^2)^{-1/2}$ to connect the branch points at $p = \pm ia$ along the imaginary axis, closing the contour for the inversion integral with an infinite semicircle in $p_r < 0$, then deforming the contour to a dumbbell-shaped figure consisting of the two vertical sides of the cut plus small circles around the branch points, and invoking the integral representation (4.1.6) for the Bessel function J_0.

2.15   Show that

$$\int_0^\infty (1 + x^2)^{-1} e^{-xt} \, dx \sim \sum_{n=0}^\infty (-)^n (2n)! \, t^{-2n-1}.$$

2.16   Use the integral representation

$$K_0(t) = \int_0^\infty e^{-t \cosh \theta} \, d\theta$$

for the modified Bessel function K_0 to obtain the asymptotic approximation

$$K_0(t) \sim \left(\frac{\pi}{2t}\right)^{1/2} e^{-t} \left[1 - \frac{1}{8t} + \frac{1^2 \cdot 3^2}{2!(8t)^2} + \cdots\right].$$

2.17   A uniform bar of unit cross section is at rest and unstressed for $t < 0$ and has free ends at both $x = 0$ and $x = l$. When $t = 0$, a force P is applied at $x = l$. Show that the subsequent displacement of any section initially at x is given by

$$y(x, t) = \frac{Pt^2}{2m} + \frac{2Pl}{\pi^2 E} \sum_{n=1}^\infty (-)^n n^{-2} \left(1 - \cos \frac{n\pi ct}{l}\right) \cos \frac{n\pi x}{l},$$

where $c^2 = E/\rho$, $m = \rho l$, and ρ and E denote density and Young's modulus. (*Note:* The Laplace transform of y has a triple pole at the origin.)   Obtain also an expression for the motion in terms of traveling waves.

2.18   Let v_0 be a function of t in the problem of Section 2.9. Use the convolution theorem to obtain the solution

$$v(x, t) = \frac{1}{2} (\pi\kappa)^{-1/2} x \int_0^t v_0(\tau)(t - \tau)^{-3/2} \exp\left[\frac{-x^2}{4\kappa(t - \tau)}\right] d\tau \,.$$

2.19   Derive (2.7.14) and use the result to obtain the asymptotic solution (2.9.11) by expanding (2.9.6) in ascending powers of p.

2.20   A slab of infinite area extends from $x = 0$ to $x = l$ and is initially at zero temperature. The temperature of the face at $x = 0$ is raised to $v = v_0$ when $t = 0$; radiation takes place at the other face in accordance with

$$v_x + hv = 0 \qquad \text{at } x = l.$$

Show that the subsequent temperature is given by

$$v(x, t) = 2v_0 \sum_k k^{-1}[h + (k^2 + h^2)l]^{-1}(k^2 + h^2)[1 - \exp(-k^2\kappa t)] \sin kx,$$

where the k are determined by

$$k \cot kl = -h.$$

★2.21　A solid is bounded internally by the cylinder $r = a$ and extends to infinity. The initial temperature is zero, and the surface is kept at a constant temperature v_0. The temperature in the solid at $t > 0$ is denoted by v and satisfies

$$v_t = \kappa(v_{rr} + r^{-1}v_r).$$

Show that the Laplace transform of v is given by

$$V(p) = \frac{v_0 K_0(qr)}{p K_0(qa)}, \qquad \text{where} \qquad q^2 = \frac{p}{\kappa},$$

and K_0 is a modified Bessel function. Show that v is given by

$$\frac{v}{v_0} = 1 + \frac{2}{\pi} \int_0^\infty \exp(-\kappa u^2 t) \frac{[J_0(ur)Y_0(ua) - J_0(ua)Y_0(ur)]}{[J_0^2(ua) + Y_0^2(ua)]} u^{-1} \, du.$$

Note:

$$K_0(xe^{\pm i\pi/2}) = (\mp i\pi/2)[J_0(x) \mp iY_0(x)] \qquad (x > 0).$$

★2.22　A uniform bar of length l, cross-sectional area A, total weight w_0, and compression-wave speed c is hanging vertically from a fixed point and stretched (in static equilibrium) under its own weight. A concentrated weight w is suddenly attached to the lower end of the bar at $t = 0$. Show that the stress of the fixed end is given by

$$\sigma = \frac{w_0}{A} + \frac{w}{A} \mathscr{L}^{-1}\left[p^{-1}\left(\cosh \frac{pl}{c} + \frac{w}{w_0} \frac{pl}{c} \sinh \frac{pl}{c} \right)^{-1} \right].$$

If $w = w_0$, show that the time at which the stress achieves its first maximum is given by

$$\frac{ct}{l} = 3 + \frac{1}{2}(1 + e^{-2}).$$

★2.23　A uniform string of length $2l$ and line density ρ has a particle of mass m attached to its midpoint and is stretched to tension ρc^2 between the fixed points $x = \pm l$. At $t = 0$, when the string is straight and at rest, the particle is set in motion by a transverse impulse I. Show that its subsequent displacement is given by

$$\frac{2lI}{mc} \sum_{n=1}^\infty \alpha_n^{-1}(1 + k \csc^2 \alpha_n)^{-1} \sin \frac{\alpha_n ct}{l},$$

where $k = 2\rho l/m$ and α_n, $n = 1, 2, \ldots$, are the positive roots of $k \cot \alpha = \alpha$.

*2.24   A constant, transverse force of w_0 per unit length is applied suddenly to a circular wire in a perfect (but compressible) fluid at $t = 0$. Prove that the subsequent velocity of the wire has the following representations

$$\frac{\pi \rho c a}{w_0} v(t) = t + \int_0^\infty \frac{e^{-ut} \, du}{u^2 [K_1^2(u) + \pi^2 I_1^2(u)]}$$

$$= 1 + \tfrac{1}{2}t + \tfrac{3}{16}t^2 + \cdots \qquad (t \to 0)$$

$$\sim t + t^{-1} + t^{-3}[\ln(4t) - \gamma - \tfrac{1}{2}] + \cdots \qquad (t \to \infty),$$

where c = velocity of sound, a = cylinder radius, $t = (c/a)$(time). Sound waves in the fluid are governed by the wave equation,

$$\phi_{rr} + \frac{1}{r} \phi_r + \frac{1}{r^2} \phi_{\theta\theta} = \phi_{tt},$$

where $\mathbf{v} = \nabla \phi$, and ar and θ are cylindrical polar coordinates.

Notes: (a) $K_1(s)$ has no zeros.

(b) $K_1(ue^{\pm i\pi}) = -K_1(u) \mp i\pi I_1(u)$.

2.25   Derive the results for the Mellin transform tabulated in the last two columns of Table 2.3 (cf. Exercise 1.1).

3 FOURIER TRANSFORMS

3.1 Introduction

We consider in this chapter the Fourier-transform pairs defined by (1.3.1), (1.3.5), and (1.3.6); see also Table 2.3. We designate the complex Fourier transform of (1.3.1) simply as the Fourier transform, and the transforms of (1.3.5) and (1.3.6) as the Fourier-cosine and Fourier-sine transforms, respectively.

We have seen that the Laplace transformation is especially suited to initial-value problems in that the transform of the nth derivative incorporates the initial values of the first $n - 1$ derivatives. The Fourier transformation, on the other hand, appears to best advantage in boundary-value problems associated with semi-infinite or infinite domains, with the appropriate selection depending on the boundary conditions and/or symmetry considerations.

3.2 Transforms of derivatives

We assume, in applying the Fourier transform to the solution of an nth-order differential equation, that $f(x)$ and its first $n - 1$ derivatives vanish at $x = \pm\infty$. Replacing $f(x)$ by its nth derivative, $f^{(n)}(x)$, in

$$F(k) = \mathscr{F}\{f(x)\} \equiv \int_{-\infty}^{\infty} f(x)e^{-ikx}\,dx \qquad (3.2.1)$$

and integrating n times by parts, we obtain

$$\mathscr{F}\{f^{(n)}(x)\} = (ik)^n F(k). \qquad (3.2.2)$$

43

An analogous result holds for the nth integral of $f(x)$ from either $x = +\infty$ or $x = -\infty$; in particular,

$$\mathscr{F}^{-1}\{(ik)^{-1}F(k)\} = \int_{\pm\infty}^{x} f(\xi)\,d\xi \qquad (3.2.3)$$

if and only if $\int_{-\infty}^{\infty} f(\xi)\,d\xi = 0$.

Similar results hold for the Fourier-cosine and Fourier-sine transforms,

$$F_c(k) = \mathscr{F}_c\{f(x)\} \equiv \int_0^\infty f(x)\cos kx\,dx \qquad (3.2.4)$$

and

$$F_s(k) = \mathscr{F}_s\{f(x)\} \equiv \int_0^\infty f(x)\sin kx\,dx. \qquad (3.2.5)$$

In particular,

$$\mathscr{F}_c\{f''(x)\} = -k^2 F_c(k) - f'(0) \qquad (3.2.6)$$

and

$$\mathscr{F}_s\{f''(x)\} = -k^2 F_s(k) + kf(0) \qquad (3.2.7)$$

if and only if both $f(x)$ and $f'(x)$ vanish at $x = \infty$.

Analogous results may be established for the cosine and sine transforms of higher derivatives of even order, but the cosine (sine) transform of an odd derivative involves the sine (cosine) transform of the original function. Thus, as implied directly by their trigonometric kernels, these transforms are intrinsically suited to differential equations having only even derivatives with respect to the variable in question. Moreover, the cosine (sine) transform of such a differential equation incorporates only the values of the odd (even) derivatives at $x = 0$; the values of other derivatives at $x = 0$ could be incorporated as constants to be determined, but the most satisfactory applications are those in which the unincorporated boundary conditions are null conditions at $x = \infty$.

The complex Fourier transform, on the other hand, may be applied to all derivatives, but it incorporates no boundary values and therefore arises naturally only for infinite domains; to be sure, it may be applied to semi-infinite domains, but then it becomes essentially a Laplace transform.

The cosine or sine transform also may be advantageously applied to an infinite domain if $f(x)$ is an even or odd function of x, respectively, in which case $f'(0)$ or $f(0)$ vanishes in consequence of symmetry.

3.3 Operational theorems

Many of the operational theorems for the Laplace transform (see Table 2.2) have analogous counterparts for the Fourier transform. We note here the shifting theorem,

$$\mathscr{F}\{f(x - a)\} = e^{-ika}F(k) \tag{3.3.1a}$$

or

$$\mathscr{F}^{-1}\{e^{-ika}F(k)\} = f(x - a), \tag{3.3.1b}$$

where a is real (but not necessarily positive), and the convolution theorem,

$$\mathscr{F}^{-1}\{F_1(k)F_2(k)\} = \frac{1}{2\pi} \int_{-\infty}^{\infty} F_1(k)F_2(k)e^{ikx}\, dk$$

$$= \int_{-\infty}^{\infty} f_1(\xi)f_2(x - \xi)\, d\xi. \tag{3.3.2}$$

Setting $x = 0$ in (3.3.2), we obtain

$$\int_{-\infty}^{\infty} f_1(\xi)f_2(-\xi)\, d\xi = \frac{1}{2\pi} \int_{-\infty}^{\infty} F_1(k)F_2(k)\, dk \tag{3.3.3}$$

or, equivalently,

$$\int_{-\infty}^{\infty} f_1(\xi)f_2(\xi)\, d\xi = \frac{1}{2\pi} \int_{-\infty}^{\infty} F_1(k)F_2^*(k)\, dk, \tag{3.3.4}$$

where

$$F^*(k) = \frac{1}{2\pi} \int_{-\infty}^{\infty} f(x)e^{ikx} \equiv F(-k) \tag{3.3.5}$$

is the complex conjugate of $F(k)$, and k is real. Letting $f_2 = f_1 = f$ in (3.3.4), we obtain *Parseval's theorem*

$$\int_{-\infty}^{\infty} f^2(x)\, dx = \frac{1}{2\pi} \int_{-\infty}^{\infty} |F(k)|^2\, dk. \tag{3.3.6}$$

The quantity $f^2(x)$ in (3.3.6) is typically proportional to the energy in physical applications, and $|F(k)|^2$—or $|F(k)|^2/(2\pi)$—is the *power spectrum* of $f(x)$.

Many of the operational theorems for the Laplace and Fourier transforms depend on the exponential kernel that characterizes these transforms; accordingly, these theorems typically have no simple counterparts for the Fourier-cosine and Fourier-sine transforms. The most important exception is the convolution theorem for the cosine transform, namely,

$$\mathscr{F}_c^{-1}\{F_c(k)G_c(k)\} = \frac{2}{\pi} \int_0^\infty F_c(k)G_c(k) \cos kx \, dk$$

$$= \frac{1}{2} \int_0^\infty f(\xi)[g(|x - \xi|) + g(x + \xi)] \, d\xi. \quad (3.3.7)$$

Setting $x = 0$ in (3.3.7), we obtain

$$\int_0^\infty f(\xi)g(\xi) \, d\xi = \frac{2}{\pi} \int_0^\infty F_c(k)G_c(k) \, dk \quad (3.3.8)$$

and

$$\int_0^\infty f^2(\xi) \, d\xi = \frac{2}{\pi} \int_0^\infty F_c^2(k) \, dk \quad (3.3.9)$$

as the analogs of (3.3.4) and (3.3.6). No such analogs exist for the Fourier-sine transform, although there are related theorems that involve both the Fourier-cosine and Fourier-sine transforms [Sneddon (1951), Section 3.6].

3.4 Initial-value problem for one-dimensional wave equation†

A classical problem in wave motion requires the solution to the wave equation

$$c^2\phi_{xx} = \phi_{tt} \quad (3.4.1)$$

for the initial values

$$\phi = f(x) \quad \text{and} \quad \phi_t = g(x) \quad \text{on} -\infty < x < \infty. \quad (3.4.2)$$

We may suppose $f(x)$ and $g(x)$ to be the initial displacement and velocity of an infinitely long string.

Taking the Fourier transform of (3.4.1, 2) with respect to x and invoking (3.2.2), we obtain

$$\Phi_{tt} + (kc)^2\Phi = 0, \quad (3.4.3)$$

$$\Phi = F(k), \quad \text{and} \quad \Phi_t = G(k) \quad \text{at } t = 0. \quad (3.4.4)$$

The solution of (3.4.3) is a linear combination of either $\cos kct$ and $\sin kct$ or $\exp(ikct)$ and $\exp(-ikct)$. Determining the coefficients in these alternative solutions to satisfy (3.4.4), we obtain

$$\Phi = F(k) \cos(kct) + (kc)^{-1}G(k) \sin(kct) \quad (3.4.5a)$$

$$= \tfrac{1}{2}[F(k) + (ikc)^{-1}G(k)]e^{ikct} + \tfrac{1}{2}[F(k) - (ikc)^{-1}G(k)]e^{-ikct}. \quad (3.4.5b)$$

---

†See Morse (1948) for further examples of this type.

Invoking the shifting theorem, (3.3.1b) with $a = \pm ct$, and (3.2.3), we invert (3.4.5b) to obtain the required solution in the form

$$\phi = \tfrac{1}{2}[f(x + ct) + f(x - ct)] + \tfrac{1}{2}c^{-1}\int_{x-ct}^{x+ct} g(\xi)\, d\xi. \qquad (3.4.6)$$

3.5 Heat conduction in a semi-infinite solid

The Fourier-cosine and Fourier-sine transforms are less flexible than the Laplace transform when applied to a semi-infinite domain; nevertheless, they may offer distinct advantages. We illustrate this last assertion for the heat-conduction problem of Section 2.9, although it should be emphasized that we are comparing the application of the Laplace transform relative to t with the Fourier transform relative to x; the Laplace transform is not well suited to x, nor is the Fourier transform well suited to t.

It is evident from (3.2.7) and the discussion in Section 3.2 that the Fourier-sine transform is well suited to the second-order differential equation (2.9.1) and the boundary conditions of (2.9.3). Designating the Fourier-sine transform of $v(x, t)$ by V and transforming (2.9.1)–(2.9.3), we obtain

$$V_t + \kappa k^2 V = \kappa k v_0 \qquad (3.5.1)$$

and

$$V = 0 \qquad \text{at } t = 0. \qquad (3.5.2)$$

The solution of (3.5.1) is a linear combination of the particular solution v_0/k and the complementary solution $\exp(-k^2\kappa t)$. Determining the coefficient of the exponential to satisfy (3.5.2), we obtain

$$V = \frac{v_0}{k}[1 - \exp(-k^2\kappa t)]. \qquad (3.5.3)$$

(This last solution could have been obtained through a Laplace transformation with respect to t.) Substituting (3.5.3) into the inversion formula (1.3.6b), we obtain

$$\frac{v}{v_0} = \frac{2}{\pi}\int_0^\infty k^{-1}[1 - \exp(-k^2\kappa t)]\sin kx\, dk. \qquad (3.5.4)$$

Invoking the known result

$$\int_0^\infty k^{-1}\sin kx\, dk = \frac{1}{2}\pi \qquad (x > 0), \qquad (3.5.5)$$

we find that (3.5.4) is equivalent to (2.9.12), and hence also to (2.9.7). Alternatively, we may invert (3.5.4) with the aid of the transform pair in EMOT Section 2.4(21),

$$\mathscr{F}_s^{-1}\{k^{-1}\exp(-ak^2\} = \operatorname{erf}(\tfrac{1}{2}a^{-1/2}x). \tag{3.5.6}$$

3.6 Two-dimensional surface-wave generation†

We illustrate the simultaneous application of integral transformations with respect to space and time variables by considering the development of two-dimensional gravity waves on a semi-infinite body of water from an initial displacement of the free surface. We regard the water as incompressible, frictionless, and initially at rest; it then follows from the known laws of hydrodynamics that the velocity and gauge pressure, say $v(x, z, t)$ and $\tilde{p}(x, z, t)$, at a given point in the fluid may be derived from a velocity potential $\phi(x, z, t)$ according to

$$\mathbf{v} = \nabla\phi \tag{3.6.1}$$

and

$$\tilde{p} = -\rho(\phi_t + gz + \tfrac{1}{2}v^2), \tag{3.6.2}$$

where ϕ satisfies Laplace's equation in two dimensions,

$$\phi_{xx} + \phi_{zz} = 0, \tag{3.6.3}$$

ρ is the density of the water, g is the gravitational acceleration, and z is measured vertically upwards from the free surface.

Let $\zeta(x, t)$ be the elevation of the free surface relative to its equilibrium position, $z = 0$. We assume that this displacement is sufficiently small to justify the neglect of all terms of second order in its amplitude—that is to say, we *linearize* the equations of motion. This assumption permits the neglect of the second-order term $\tfrac{1}{2}\rho v^2$ in (3.6.2) and the evaluation of ϕ and its derivatives at $z = 0$, rather than $z = \zeta$, in the free-surface boundary conditions.‡ The kinematical boundary condition on the vertical velocity at this surface then is, from the definitions of ϕ and ζ,

$$\phi_z = \zeta_t \quad \text{at } z \doteq 0. \tag{3.6.4a}$$

---

†See Lamb (1932). Sections 238–241 or Sneddon (1951), Sections 32.1–32.4 for a more complete exposition and references (note that Lamb's velocity potential is opposite in sign to that used here). This problem and that of Section 4.3 were originally posed and solved independently by Cauchy and Poisson, whence it is known as the (two-dimensional) Cauchy–Poisson problem.

‡This approximation assumes that ϕ and its first derivatives can be expanded in powers of ζ about $z = 0$.

The dynamical boundary condition, corresponding to the requirement $\tilde{p} = 0$ at the free surface (we neglect the aerodynamic reaction as small compared with the hydrodynamic forces), is

$$\tilde{p} \doteq -\rho(\phi_t + g\zeta) = 0 \qquad \text{at } z \doteq 0. \tag{3.6.4b}$$

We complete the statement of the boundary conditions by invoking the *finiteness* conditions

$$|\phi| < \infty \qquad \text{as } x \rightarrow \pm\infty \text{ or } z \rightarrow -\infty \tag{3.6.5a}$$

and

$$|\zeta| < \infty \qquad \text{as } x \rightarrow \pm\infty. \tag{3.6.5b}$$

Prescribing the initial displacement $\zeta_0(x)$ and invoking the assumption that the water is initially at rest, we obtain the initial conditions

$$\phi = 0 \qquad \text{and} \qquad \zeta = \zeta_0(x) \qquad \text{at } t = 0. \tag{3.6.6}$$

We attack the mathematical problem posed by (3.6.3)–(3.6.6) by invoking a Fourier transformation with respect to x and a Laplace transformation with respect to t. Let

$$\Phi(k, z, p) = \mathscr{L}\mathscr{F}\phi, \qquad Z(k, p) = \mathscr{L}\mathscr{F}\zeta, \qquad \text{and} \qquad Z_0(k) = \mathscr{F}\zeta_0 \tag{3.6.7}$$

denote the required transforms, where $\mathscr{F}$ implies Fourier transformation with respect to $x \, (-\infty < x < \infty)$, and $\mathscr{L}$ implies Laplace transformation with respect to $t \, (0 < t < \infty)$. Transforming (3.6.3) with the aid of (3.2.2), we obtain

$$\Phi_{zz} - k^2\Phi = 0. \tag{3.6.8}$$

Transforming (3.6.4a, b) and invoking (3.6.6), we obtain

$$-\Phi_z + pZ = Z_0 \qquad \text{and} \qquad p\Phi + gZ = 0 \qquad \text{at } z = 0. \tag{3.6.9}$$

The solutions of (3.6.8) are proportional to $\exp(\pm kz)$, where the upper (lower) sign must be chosen for $k > (<) 0$ in order to satisfy the finiteness condition (3.6.5a) as $z \rightarrow -\infty$; accordingly, the required solution has the form

$$\Phi = F(k, p)e^{|k|z}. \tag{3.6.10}$$

Substituting (3.6.10) into (3.6.9), eliminating Z between the resulting equations, and solving for F, we obtain

$$\Phi = -g(p^2 + g|k|)^{-1}Z_0(k)e^{|k|z}. \tag{3.6.11}$$

Taking the inverse-Laplace transform of (3.6.11) with the aid of T2.1.4 and writing out the inverse-Fourier transform with the aid of (1.3.1b), we obtain

$$\phi = -(2\pi)^{-1}g^{1/2}\int_{-\infty}^{\infty} |k|^{-1/2}Z_0(k)e^{|k|z+ikx} \sin[\omega(k)t] \, dk, \quad (3.6.12)$$

where

$$\omega(k) = \omega(-k) = (g|k|)^{1/2} \qquad (3.6.13)$$

is the angular frequency of a gravity wave of wave number k.

Substituting (3.6.12) into (3.6.4b), we obtain

$$\zeta = -g^{-1}\phi_t|_{z=0} = \frac{1}{2\pi}\int_{-\infty}^{\infty} Z_0(k)e^{ikx} \cos \omega t \, dk \qquad (3.6.14a)$$

$$= \frac{1}{\pi}\mathscr{R}\int_0^{\infty} Z_0(k)e^{ikx} \cos \omega t \, dk \qquad (3.6.14b)$$

$$= \frac{1}{2\pi}\mathscr{R}\int_0^{\infty} Z_0(k)[e^{i(kx-\omega t)} + e^{i(kx+\omega t)}] \, dk, \qquad (3.6.14c)$$

where (3.6.14b, c) follow from (3.6.14a) by virtue of the identities $Z_0(-k) = Z_0^*(k)$, $\omega(-k) = \omega(k)$, and the exponential representation of the cosine. The first and second terms in the integrand of (3.6.14c) represent waves traveling in the directions of increasing and decreasing x, respectively, with the phase velocity ω/k; the corresponding integrals represent superpositions of such waves over the wave-number spectrum, $0 < k < \infty$.

The free-surface displacement can be expressed in terms of tabulated functions (Fresnel integrals) in the special case of a delta-function displacement (this development is given by Lamb), and the result can be generalized with the aid of the convolution theorem. However, it suffices for many purposes and is, in any event, more instructive to consider the asymptotic representation of (3.6.14c) by the method of stationary phase, which we now proceed to develop.

3.7 *The method of stationary phase*

We consider the asymptotic behavior, as $t \to \infty$, of the integral

$$I = \int_a^b A(k)e^{i\phi(k)} \, dk, \qquad (3.7.1a)$$

where

$$\phi(k) = kx - \omega(k)t \qquad (x > 0) \qquad (3.7.1b)$$

is the phase and $A(k)$ the amplitude density (either real or complex) of plane waves in the wave-number spectrum $a < k < b$. The form of $\phi(k)$

displayed in (3.7.1b) is especially convenient for problems in wave propagation, but it would be equally general to let $\phi = tf(k)$, which reduces to (3.7.1b) for $f = k(x/t) - \omega(k)$.

The integrand of (3.7.1), regarded as a function of k, oscillates with increasing rapidity as $t \to \infty$, in consequence of which the contributions to I of adjacent portions of the integrand cancel one another except in the neighborhoods of the end points (there are no contributions from just below $k = a$ to cancel those from just above $k = a$ and conversely for k near b) and those points, if any, at which $\phi(k)$ is stationary, say

$$\phi'(k_s) = 0 \qquad (a < k_s < b). \tag{3.7.2}$$

The points determined by (3.7.2) are known as *points of stationary phase*, and here and subsequently the subscript s implies evaluation at $k = k_s$. We consider separately the special cases $k_s \to a$ or $k_s \to b$.

Let us suppose that I has one and only one point of stationary phase; if it has n such points, it may be subdivided into n integrals, each of which has only one such point. Expanding $A(k)$ and $\phi(k)$ in Taylor series about $k = k_s$, we obtain

$$I = \int_a^b [A_s + A_s'(k - k_s) + \tfrac{1}{2}A_s''(k - k_s)^2 + \cdots]$$

$$\cdot \exp\{i[\phi_s + \tfrac{1}{2}\phi_s''(k - k_s)^2 + \tfrac{1}{6}\phi_s'''(k - k_s)^3 + \cdots]\}\, dk. \tag{3.7.3}$$

Introducing the change of variable $k = k_s + \varepsilon u$, where

$$\varepsilon = \left(\frac{2}{|\phi_s''|}\right)^{1/2} = \left(\frac{2}{|t\omega_s''|}\right)^{1/2},$$

we obtain

$$I = \varepsilon \int_{-(k_s - a)/\varepsilon}^{(b - k_s)/\varepsilon} [A_s + \varepsilon A_s' u + \tfrac{1}{2}\varepsilon^2 A_s'' u^2 + \cdots]$$

$$\cdot \exp\left\{i\left[\phi_s + u^2\, \mathrm{sgn}\, \phi_s'' + \tfrac{1}{3}\varepsilon\left(\frac{\phi_s'''}{|\phi_s''|}\right)u^3 + \cdots\right]\right\} du, \tag{3.7.4}$$

where sgn implies *the sign of* (sgn x is known as the *signum* function of x). Letting $\varepsilon \to 0$ $(t \to \infty)$ in (3.7.4) and invoking the known integral

$$\int_{-\infty}^{\infty} \exp(\pm iu^2)\, du = \pi^{1/2} \exp(\pm\tfrac{1}{4}i\pi),$$

we obtain

$$I \sim \left(\frac{2\pi}{|\phi_s''|}\right)^{1/2} A_s \exp[i(\phi_s + \tfrac{1}{4}\pi\, \mathrm{sgn}\, \phi_s'')] \qquad (t \to \infty). \tag{3.7.5}$$

Repeating the argument on the supposition that either $k_s = a$ or $k_s = b$, we infer from (3.7.4) that the asymptotic limits of integration for u become either $(0, \infty)$ or $(-\infty, 0)$, rather than $(-\infty, \infty)$, in consequence of which the right-hand side of (3.7.5) must be divided by 2. [It is evident from this result that the asymptotic approximation (3.7.5) is not uniformly valid as k_s approaches either a or b.]

The approximation (3.7.5) obviously fails if $\phi_s'' = 0$. This important special case and others of less importance are discussed in the monographs of Copson (1965) and Erdélyi (1956). The approximation may be extended to complex $\phi(k)$ [such as would arise in approximating the integral of (3.6.12) if z, as well as x and t, is large], in which case it appears as Riemann's *saddle-point approximation* [Copson (1965, Section 36)], a special case of Debye's *method of steepest descent* [Copson (1965, Section 29)].

The asymptotic behavior of (3.7.1) may be determined through integration by parts if $\phi'(k)$ does not vanish in $a \leq k \leq b$. The result is

$$I \sim \frac{iA(a)}{\phi'(a)} \exp[i\phi(a)] - \frac{iA(b)}{\phi'(b)} \exp[i\phi(b)] \qquad (t \to \infty). \qquad (3.7.6)$$

The right-hand sides of (3.7.5) and (3.7.6) may be superimposed to obtain the leading terms in the complete asymptotic expansion of I provided that $k_s \neq a$ or b. More general cases are discussed by Erdélyi (1956).

Returning now to the problem of Section 3.6, we observe that only the first of the two exponentials in (3.6.14c) has a point of stationary phase for $x > 0$; accordingly, this exponential dominates the asymptotic approximation to the integral, such that

$$\zeta \sim (2\pi)^{-1}\mathscr{R}\int_0^\infty Z_0(k)e^{i\phi(k)}\,dk \qquad (x > 0, t \to \infty), \qquad (3.7.7)$$

where ϕ is given by (3.7.1b). Invoking (3.6.13) and (3.7.2), we obtain

$$\phi'(k_s) = x - \omega'(k_s)t = x - \frac{1}{2}(g/k_s)^{1/2}t = 0, \qquad (3.7.8)$$

which implies

$$k_s = \frac{gt^2}{4x^2}, \qquad \phi_s = -\frac{gt^2}{4x^2}, \qquad \text{and} \qquad \phi_s'' = \frac{2x^3}{gt^2} \qquad (3.7.9)$$

Substituting (3.7.9) into (3.7.5), setting $A(k) = Z_0(k)/(2\pi)$, and taking the real part of the result, we obtain

$$\zeta \sim \frac{1}{2}\pi^{-1/2}g^{1/2}x^{-3/2}t\,\mathcal{R}\left\{Z_0\left(\frac{gt^2}{4x^2}\right)\exp\left[-\frac{1}{4}i\left(\frac{gt^2}{x}+\pi\right)\right]\right\}$$

$$(x > 0, t \to \infty). \quad (3.7.10)$$

A consideration of the neglected terms in the stationary-phase approximation implies that (3.7.10) is valid for $gt^2 \gg 4x$. If $x < 0$, only the second exponential in (3.6.14c) has a point of stationary phase, and a repetition of the preceding argument with $\phi = -k|x| + \omega(k)t$ yields

$$\zeta \sim \frac{1}{2}\pi^{-1/2}g^{1/2}|x|^{-3/2}t\,\mathcal{R}\left\{Z_0\left(\frac{gt^2}{4x^2}\right)\exp\left[\frac{1}{4}i\left(\frac{gt^2}{|x|}+\pi\right)\right]\right\}$$

$$(x < 0, t \to \infty). \quad (3.7.11)$$

We may render (3.7.11) valid for $x > 0$ simply by multiplying the exponent by $-\operatorname{sgn} x$.

3.8 Fourier transforms in two or more dimensions

We may extend the Fourier-transform pair of (1.3.1a, b) to a function of two variables to obtain

$$F(k_1, k_2) = \int_{-\infty}^{\infty}\int_{-\infty}^{\infty} f(x, y)\exp[-i(k_1 x + k_2 y)]\,dx\,dy \quad (3.8.1a)$$

and

$$f(x, y) = \frac{1}{4\pi^2}\int_{-\infty}^{\infty}\int_{-\infty}^{\infty} F(k_1, k_2)\exp[i(k_1 x + k_2 y)]\,dk_1\,dk_2. \quad (3.8.1b)$$

More generally, letting $\mathbf{r}$ denote a vector having the Cartesian components $x_1, x_2, \ldots, x_n$ in an n-dimensional space and $\mathbf{k}$ a similar vector in the wavenumber space $k_1, k_2, \ldots, k_n$, we obtain

$$F(\mathbf{k}) = \int_{-\infty}^{\infty}\cdots\int_{-\infty}^{\infty} f(\mathbf{r})e^{-i\mathbf{k}\cdot\mathbf{r}}\,dx_1\cdots dx_n \quad (3.8.2a)$$

and

$$f(\mathbf{r}) = (2\pi)^{-n}\int_{-\infty}^{\infty}\cdots\int_{-\infty}^{\infty} F(\mathbf{k})e^{i\mathbf{k}\cdot\mathbf{r}}\,dk_1\cdots dk_n. \quad (3.8.2b)$$

Further transform pairs may be obtained from (3.8.1a, b) and (3.8.2a, b) by coordinate transformations. In particular, the transformation of (3.8.1a, b) to polar coordinates leads to the Hankel transform (1.1.6) and its inverse. We give the derivation in Section 4.1.

The Laplace-transform pair of (1.4.3) may be extended to functions of two variables through transformation of the Fourier-transform pair of (3.8.1); see Ditkin–Prudnikov (1962) and Voelker–Doetsch (1950).

EXERCISES

3.1 Determine the Fourier transform of Heaviside's step function, $H(x)$, by (a) allowing k to be complex (state the necessary restriction on k_i) and (b) transforming $e^{-ax}H(x)$ and considering the limit $a \to 0+$.

Answer: (a) $\mathcal{F}\{H(x)\} = (ik)^{-1}$ $(k_i < 0)$.

        (b) $\mathcal{F}\{e^{-ax}H(x)\} = (a + ik)^{-1}$ $(a \to 0+)$.

3.2 Determine the Fourier transform of the *signum function*,

$$\text{sgn } x = \pm 1 \qquad (x \gtrless 0),$$

by transforming $e^{-a|x|} \text{sgn } x$ and considering the limit $a \to 0+$. Compare the result with those of Exercise 3.1.a, b and discuss the seeming paradox. Can the difficulty be resolved by permitting k to be complex?

Answer: $\mathcal{F}\{e^{-a|x|} \text{sgn } x\} = -2ik(a^2 + k^2)^{-1}$. The limiting result, $2(ik)^{-1}$, *appears* to be the Fourier transform of $2H(x)$ and cannot be rendered valid by permitting k to be complex.

3.3 Show that the Fourier transform of $f(x) = x^{-1} \sin mx$ is $F(k) = \pi H(m - |k|)$, where $m > 0$ and $k_i = 0$.

3.4 Show that the inverse Fourier transform of

$$F(k) = \beta[(\alpha + ik)^2 + \beta^2]^{-1}$$

is

$$f(x) = H(x)e^{-\alpha x} \sin \beta x$$

by an appropriate deformation of the path of the inversion integral in the complex-k plane.

3.5 Use the Fourier-sine transform to solve the following problem in potential theory:

$$\phi_{xx} + \phi_{yy} = 0 \qquad (x > 0, \, y > 0),$$

$$\phi = 1 \text{ on } x = 0, \, y > 0, \qquad \phi = 0 \text{ on } x > 0, \, y = 0,$$

and

$$\nabla\phi \to 0 \qquad \text{as } x^2 + y^2 \to \infty.$$

Answer:

$$\phi = \frac{2}{\pi} \int_0^\infty k^{-1}(1 - e^{-ky}) \sin kx \, dk = \frac{2}{\pi} \tan^{-1} \frac{y}{x}.$$

3.6 The transverse oscillations of a semi-infinite string, $x > 0$, are governed by the wave equation, (3.4.1). The string is originally at rest, and the end at $x = 0$

is subjected to the displacement $f(t)$ for $t > 0$. Obtain the classical solution

$$y(x, t) = f\left(t - \frac{x}{c}\right) H\left(t - \frac{x}{c}\right)$$

by invoking a Fourier-sine transformation with respect to x and a Laplace transformation with respect to t.

*3.7 The preceding result may be inferred directly from the known properties of the wave equation—in particular, the fact that the wave propagation is non-dispersive. In contrast, wave propagation along an ideal beam, which is governed by

$$a^2 y_{xxxx} + y_{tt} = 0 \qquad \left(a^2 = \frac{EI}{m}\right),$$

where EI is the bending stiffness and m is the mass per unit length, *is* dispersive. Consider a semi-infinite beam that is initially at rest and for which the freely hinged end has the prescribed motion $f(t)$, so that

$$y = f(t), \qquad y_{xx} = 0 \qquad \text{at } x = 0, \ t > 0.$$

Show that the Laplace transform of $y(x, t)$ is given by

$$\mathscr{L} y(x, t) = [\mathscr{L} f(t)] \exp\left[-\left(\frac{p}{2a}\right)^{1/2} \cdot x\right] \cos\left[\left(\frac{p}{2a}\right)^{1/2} x\right]$$

and invert this result to obtain

$$y(x, t) = (8\pi a)^{-1/2} x \int_0^t f(\tau)(t - \tau)^{-3/2} \{\sin[\tfrac{1}{4} a^{-1} x^2 (t - \tau)^{-1}]$$
$$+ \cos[\tfrac{1}{4} a^{-1} x^2 (t - \tau)^{-1}]\} \, d\tau.$$

3.8 The initial temperature in an infinite medium is given by $v_0 = f(x)$. Use a Fourier transformation to obtain the solution of (2.9.1) in the form

$$v(x, t) = (4\pi\kappa t)^{-1/2} \int_{-\infty}^{\infty} f(\xi) \exp[-(x - \xi)^2 (4\kappa t)^{-1}] \, d\xi$$
$$= \pi^{-1/2} \int_{-\infty}^{\infty} f[x + 2(\kappa t)^{1/2} \eta] \exp(-\eta^2) \, d\eta.$$

3.9 Obtain the counterpart of (3.6.12) for $\zeta_0(x) = \delta(x)$ by invoking symmetry considerations in the original statement of the problem and using a Fourier-cosine transform. [N.B. $\mathscr{L}\delta(t) = 1$, but $\mathscr{F}_c\delta(x) = \tfrac{1}{2}$; justify the latter result.]

3.10 Repeat the solution of Section 3.6 for $\zeta_0(x) = 0$ and

$$\phi(x, 0, 0) = \frac{I}{\rho} \delta(x),$$

corresponding to the application of a concentrated impulse I at $t = 0$.

Answer:

$$\rho\phi = \frac{I}{\pi} \int_0^\infty e^{kz} \cos(kx) \cos[(gk)^{1/2}t]\, dk.$$

3.11 Use the method of stationary phase to obtain the asymptotic form of the free-surface displacement in Exercise 3.10.

4 HANKEL TRANSFORMS

4.1 Introduction

The Hankel-transform pair,

$$F_n(k) = \mathscr{H}_n\{f(r)\} \equiv \int_0^\infty f(r)J_n(kr)r\, dr \qquad (4.1.1a)$$

and

$$f(r) = \mathscr{H}_n^{-1}\{F_n(k)\} \equiv \int_0^\infty F_n(k)J_n(kr)k\, dk, \qquad (4.1.1b)$$

arises naturally in connection with the differential operator

$$\Delta_n \equiv \left(\frac{\partial}{\partial r}\right)^2 + \frac{1}{r}\frac{\partial}{\partial r} - \left(\frac{n}{r}\right)^2 = \frac{1}{r}\frac{\partial}{\partial r}\frac{r\partial}{\partial r} - \left(\frac{n}{r}\right)^2, \qquad (4.1.2)$$

which is derived from the Laplacian operator

$$\nabla^2 = \left(\frac{\partial}{\partial r}\right)^2 + \frac{1}{r}\frac{\partial}{\partial r} + \frac{1}{r^2}\left(\frac{\partial}{\partial \theta}\right)^2 + \left(\frac{\partial}{\partial z}\right)^2 \qquad (4.1.3)$$

after separation of variables in cylindrical polar coordinates (r, θ, z).

We derive (4.1.1a, b) by introducing the polar-coordinate transformations $x = r\cos\theta$, $y = r\sin\theta$, $k_1 = k\cos\alpha$, and $k_2 = k\sin\alpha$ in the two-dimensional, Fourier-transform pair of (3.8.1a, b) to obtain (after appropriate changes in functional notation)

$$F(k, \alpha) = \int_0^\infty \int_0^{2\pi} f(r, \theta)\exp[-ikr\cos(\theta - \alpha)]r\, dr\, d\theta \qquad (4.1.4a)$$

and

$$f(r, \theta) = \frac{1}{4\pi^2} \int_0^\infty \int_0^{2\pi} F(k, \alpha) \exp[ikr \cos(\theta - \alpha)]k\, dk\, d\alpha. \quad (4.1.4b)$$

[We remark that if $f(r, \theta)$ is multiplied by $\exp(-i\omega t)$, (4.1.4b) represents a packet of plane waves having the amplitude distribution $F(k, \alpha)$, the wave speeds ω/k, and wave-front normals inclined at the angles α to the x axis.] Now let us suppose that

$$f(r, \theta) = f(r)e^{in\theta}; \quad (4.1.5)$$

this is not a restrictive assumption, for we may expand $f(r, \theta)$ in a complex Fourier series, in which each term has the form (4.1.5), and then consider the series term by term. Substituting (4.1.5) into (4.1.4a) and introducing the change of variable $\varphi = \theta - \alpha + \frac{1}{2}\pi$, we obtain

$$F(k, \alpha) = \exp[in(\alpha - \tfrac{1}{2}\pi)] \int_0^\infty f(r)r\, dr \int_{-(1/2)\pi - \alpha}^{(3/2)\pi - \alpha} \exp[i(n\varphi - kr \sin \varphi)]\, d\varphi.$$

Invoking the representation [Watson (1945), Section 2.2(5)]

$$2\pi J_n(kr) = \int_{\varphi_0}^{2\pi + \varphi_0} \exp[i(n\varphi - kr \sin \varphi)]\, d\varphi \quad (4.1.6)$$

for the Bessel function J_n, choosing $\varphi_0 = \frac{1}{2}\pi - \alpha$, and identifying the integral over r with the Hankel transform of $f(r)$, as defined by (4.1.1a), we obtain

$$F(k, \alpha) = 2\pi \exp[in(\alpha - \tfrac{1}{2}\pi)]F_n(k). \quad (4.1.7)$$

Substituting (4.1.7) into (4.1.4b), introducing the change of variable $\varphi = \alpha - \theta - \frac{1}{2}\pi$, and comparing the result to (4.1.5), we obtain

$$f(r)e^{in\theta} = \frac{1}{2\pi} \int_0^\infty F_n(k)k\, dk \int_{-(1/2)\pi - \theta}^{(3/2)\pi - \theta} \exp[i(n\varphi - kr \sin \varphi)]\, d\varphi\, e^{in\theta}.$$

Invoking (4.1.6) with $\varphi_0 = -\frac{1}{2}\pi - \theta$ and cancelling $\exp(in\theta)$, we obtain (4.1.1b), thereby establishing the inverse relation between (4.1.1a, b).

Another form of the Hankel transform, which is used in EMOT, is given by

$$G_n(k) = \int_0^\infty g_n(r)J_n(kr)(kr)^{1/2}\, dr \quad (4.1.8a)$$

and

$$g_n(r) = \int_0^\infty G_n(k)J_n(kr)(kr)^{1/2}\, dk. \quad (4.1.8b)$$

This evidently can be reconciled with (4.1.1a, b) by setting $g_n = r^{1/2}f_n$ and

$G_n = k^{1/2}F_n$. As it stands, it reduces, except for the constant factor $(2/\pi)^{1/2}$, to the Fourier-sine or -cosine transform for $n = \frac{1}{2}$ or $-\frac{1}{2}$, respectively.

Returning now to the operator Δ_n of (4.1.2), we take the nth-order Hankel transform of $\Delta_n\phi$ and integrate twice by parts to obtain

$$\mathcal{H}_n\{\Delta_n\phi\} = \int_0^\infty \{r^{-1}(r\phi_r)_r - n^2r^{-2}\phi\}J_n(kr)r\,dr$$

$$= \left(r\phi_rJ_n - r\phi\frac{d}{dr}J_n\right)\Bigg|_0^\infty + \int_0^\infty \{\Delta_nJ_n(kr)\}\phi r\,dr.$$

Assuming that the partially integrated terms vanish at both limits (note that J_n vanishes like r^n as $r \to 0$ and like $r^{-1/2}$ as $r \to \infty$) and invoking Bessel's equation,

$$(\Delta_n + k^2)J_n(kr) = 0, \tag{4.1.9}$$

we obtain

$$\mathcal{H}_n\{\Delta_n\phi\} = -k^2\mathcal{H}_n\phi. \tag{4.1.10}$$

The most important, special cases of the Hankel transform correspond to $n = 0$ and $n = 1$, but we emphasize that, subject to appropriate restrictions on ϕ, the reciprocal relation implied by (4.1.1a, b) and the result (4.1.10) are valid for $\mathcal{R}n > -\frac{1}{2}$ [see Sneddon (1951), Chapter 2].

4.2 Oscillating piston

We consider an oscillating piston of radius a mounted in an infinite plane, $z = 0$, and radiating sound into a half-space, as shown in Figure 4.1. This configuration serves as a simple model of a loudspeaker.†

Let the displacement of the piston from its equilibrium position, $z = 0$, be given by

$$z = \mathcal{R}\{Ae^{i\omega t}\}, \tag{4.2.1}$$

where A is a complex amplitude, and ω is the angular frequency. We assume that $|A|$ is small compared with the other characteristic lengths defined by the problem, namely a and c/ω, where c is the velocity of sound (the wavelength corresponding to the angular frequency ω is $2\pi c/\omega$). We require a solution to the wave equation,

$$c^2\nabla^2\phi = \phi_{tt}, \tag{4.2.2}$$

---

† This problem, which was solved originally by Lord Rayleigh, is discussed in considerable detail by Morse (1948, pp. 326–338).

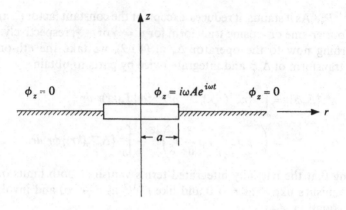

FIGURE 4.1 Oscillating piston in an infinite baffle.

for the complex velocity potential ϕ (only the real part of which is to be retained in the calculation of the actual velocity). The boundary condition is (we prescribe the piston velocity at $z = 0$, rather than the displaced position, by virtue of our assumption that $|A|$ is small)

$$\phi_z = i\omega A e^{i\omega t} H(a - r) \qquad \text{at } z = 0, \tag{4.2.3}$$

where H is Heaviside's step function. We also require ϕ to be bounded and to yield an outgoing wave at infinity (radiation condition).

We begin our construction of the solution by remarking that ϕ must be independent of θ by virtue of the fact that the boundary conditions are similarly independent. We also may replace ϕ_{tt} by $-\omega^2\phi$ by virtue of the prescribed time dependence. Invoking these results in (4.2.2), we obtain

$$\phi_{zz} + \phi_{rr} + \frac{1}{r}\phi_r + \left(\frac{\omega}{c}\right)^2 \phi = 0. \tag{4.2.4}$$

Applying the operator $\mathscr{H}_0$ to (4.2.3) and (4.2.4) and defining

$$\Phi = \mathscr{H}_0(\phi), \tag{4.2.5}$$

we obtain

$$\Phi_{zz} - \lambda^2\Phi = 0, \tag{4.2.6}$$

where

$$\lambda = \left[k^2 - \left(\frac{\omega}{c}\right)^2\right]^{1/2}, \tag{4.2.7}$$

and

$$\Phi_z\big|_{z=0} = i\omega A e^{i\omega t}\int_0^a J_0(kr)r\,dr = i\omega A a k^{-1}J_1(ka)e^{i\omega t}. \tag{4.2.8}$$

The required solution of (4.2.6) and (4.2.8) is

$$\Phi = -i\omega Aa(k\lambda)^{-1}J_1(ka)e^{i\omega t - \lambda z}, \tag{4.2.9}$$

where λ must be positive real for $k > \omega/c$ and positive imaginary for $k < \omega/c$ in consequence of the requirements that the solution be bounded and behave as an outgoing wave, as $x \to \infty$ [let $\lambda = i\mu$ for $k < \omega/c$; then (4.2.9) represents a disturbance moving in the positive-x direction with the phase speed ω/μ]. Applying the operator $\mathcal{H}_0^{-1}$, as defined by (4.1.1b), to Φ, we obtain

$$\phi = -i\omega Aa \int_0^\infty \lambda^{-1} J_1(ka)J_0(kr)e^{i\omega t - \lambda z}\,dk. \tag{4.2.10}$$

The gauge pressure on the piston is given by

$$\tilde{p} = -\rho\phi_t|_{z=0} = -\rho\omega^2 Aae^{i\omega t}\int_0^\infty \lambda^{-1} J_1(ka)J_0(kr)\,dk. \tag{4.2.11}$$

The corresponding power (which goes into acoustic radiation) is given by

$$P = \tfrac{1}{2}\mathcal{R}\left\{\left[2\pi\int_0^a \tilde{p}r\,dr\right]\left[i\omega Ae^{i\omega t}\right]^*\right\}, \tag{4.2.12}$$

where the first and second terms in square brackets are the complex force and the complex conjugate of the velocity, respectively. Substituting (4.2.11) into (4.2.12), carrying out the integration with respect to r, and taking the real part of the result (to which only that portion of the k spectrum in which λ is imaginary contributes), we obtain

$$P = \pi a^2|A|^2\rho\omega^3\int_0^{\omega/c} k^{-1}\left[\left(\frac{\omega}{c}\right)^2 - k^2\right]^{-1/2} J_1^2(ka)\,dk. \tag{4.2.13}$$

Introducing the change of variable $k = (\omega/c)\sin\alpha$, we find that the integral in (4.2.13) can be evaluated [Luke (1962), Section 13.3.2(24)] to obtain

$$P = \frac{1}{2}\pi a^2\rho\omega^2|A|^2 c\left[1 - \frac{c}{\omega a}J_1\left(\frac{2\omega a}{c}\right)\right], \tag{4.2.14}$$

a result due originally to Lord Rayleigh.

4.3 Axisymmetric surface-wave generation †

We consider the axisymmetric analog of the hydrodynamical problem of Section 3.6, namely, the development of gravity waves on a semi-infinite body of water from the initial displacement $\zeta = \zeta_0(r)$. The mathematical

---

† See Lamb (1932), Section 255 and Sneddon (1951), Section 32.6 for more detailed discussions of this problem.

development closely parallels that of Section 3.6, with x replaced by the cylindrical radius r and the Fourier-transform operator $\mathscr{F}$ replaced by the Hankel-transform operator $\mathscr{H}_0$.

Let $\phi(r, z, t)$ be the velocity potential. The velocity and gauge pressure, $v(r, z, t)$ and $\tilde{p}(r, z, t)$, then are given by (3.6.1) and (3.6.2), and ϕ satisfies Laplace's equation for axisymmetric motion,

$$\phi_{rr} + r^{-1}\phi_r + \phi_{zz} = 0, \qquad (4.3.1)$$

in place of (3.6.3). The boundary conditions on the assumption of small displacements are given by (3.6.4) and (3.6.5). The initial conditions are given by (3.6.6). Carrying out a zero-order Hankel transformation with respect to r and a Laplace transformation with respect to t, we write

$$\Phi(k, z, p) = \mathscr{L}\mathscr{H}_0\phi, \qquad Z(k, p) = \mathscr{L}\mathscr{H}_0\zeta, \qquad Z_0(k) = \mathscr{H}_0\zeta_0 \quad (4.3.2)$$

in place of (3.6.7). Transforming (4.3.1), (3.6.4), and (3.6.6), we obtain the boundary-value problem posed by (3.6.8) and (3.6.9). The required solution is given by (3.6.11) or, since k is nonnegative for the Hankel transform,

$$\Phi = -g(p^2 + gk)^{-1}Z_0(k)e^{kz}. \qquad (4.3.3)$$

The solution departs from that of Section 3.6 at this point in consequence of the geometrical differences. Inverting (4.3.3) with the aid of T2.1.4 and (4.1.1b), we obtain

$$\phi(r, z, t) = -g^{1/2}\int_0^\infty k^{1/2}Z_0(k)J_0(kr)e^{kz} \sin \omega t \, dk, \qquad (4.3.4)$$

where $\omega(k) = (gk)^{1/2}$, as in Section 3.6.

We now obtain the asymptotic approximation to the free-surface displacement with the aid of the stationary-phase approximation of Section 3.7. Setting $z = 0$ in (4.3.4) on the assumption that $Z_0(k)$ vanishes with sufficient rapidity as $k \to \infty$ to ensure the convergence of the integral and invoking the boundary condition (3.6.4b) to determine ζ, we obtain

$$\zeta(r, t) = -g^{-1}\phi_t(r, 0, t) = \int_0^\infty kZ_0(k)J_0(kr) \cos \omega t \, dk. \qquad (4.3.5)$$

Letting $r \to \infty$ and replacing J_0 by the asymptotic approximation (cf. Exercise 4.4)

$$J_0(kr) \sim \left(\frac{2}{\pi kr}\right)^{1/2} \cos(kr - \tfrac{1}{4}\pi), \qquad (4.3.6)$$

we obtain

$$\zeta(r, t) \sim \left(\frac{2}{\pi r}\right)^{1/2} \int_0^\infty k^{1/2} Z_0(k) \cos(kr - \tfrac{1}{4}\pi) \cos \omega t \, dk. \qquad (4.3.7)$$

Resolving the product of the cosines into the cosines of the sum and difference of the arguments, we obtain a representation similar to that of (3.6.14c),

$$\zeta(r, t) \sim (2\pi r)^{-1/2} \mathscr{R} \int_0^\infty k^{1/2} Z_0(k) \exp[i(\omega t - kr + \tfrac{1}{4}\pi)] \, dk, \quad (4.3.8)$$

where we have retained only that exponential which has a point of stationary phase, namely [cf. (3.7.9)],

$$k_s = \frac{gt^2}{4r^2}. \qquad (4.3.9)$$

Carrying out the stationary-phase approximation (the student should fill in the details, following the example in Section 3.7), we obtain

$$\zeta(r, t) \sim \frac{gt^2}{2^{3/2} r^3} Z_0\left(\frac{gt^2}{4r^2}\right) \cos \frac{gt^2}{4r} \qquad \left(\frac{gt^2}{4r} \to \infty\right). \qquad (4.3.10)$$

The classical solution for the Cauchy-Poisson problem is based on an initial displacement of unit volume that is concentrated at the origin, which yields $Z_0(k) = (2\pi)^{-1}$. However, actual displacements of an incompressible fluid must satisfy the constraint of zero net volume,

$$V = 2\pi \int_0^\infty \zeta_0(r) r \, dr = 0, \qquad (4.3.11)$$

which implies

$$Z_0(0) = 0. \qquad (4.3.12)$$

A displacement that does satisfy (4.3.11) and resembles the initial displacement produced by a surface explosion† or a dropped pebble is given by

$$\zeta_0(r) = d\left[1 - \left(\frac{r}{a}\right)^2\right] \exp\left[-\left(\frac{r}{a}\right)^2\right], \qquad (4.3.13)$$

which represents a cavity of depth d with a concentric lip, as shown in Figure 4.2. Invoking EMOT 8.3(5),

---

† It need scarcely be added that the motion near an explosion is too violent to justify the assumption of small displacements. Nevertheless, this assumption permits an adequate description of the motion at a sufficient distance from the explosion.

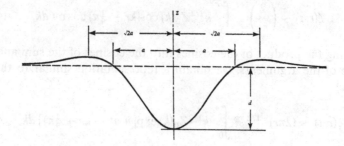

FIGURE 4.2   The cavity and lip described by (4.3.13) [from *J. Fluid Mech.*
34, 368 (1968) by courtesy of Cambridge University Press].

$$\mathscr{H}_0\left\{\left[1-\left(\frac{r}{a}\right)^2\right]\exp\left[-\left(\frac{r}{a}\right)^2\right]\right\}=\frac{1}{8}a^4k^2\exp\left[-\frac{1}{4}(ka)^2\right], \qquad (4.3.14)$$

we obtain

$$Z_0(k)=\tfrac{1}{8}da^2(ka)^2\exp\left[-\tfrac{1}{4}(ka)^2\right]. \qquad (4.3.15)$$

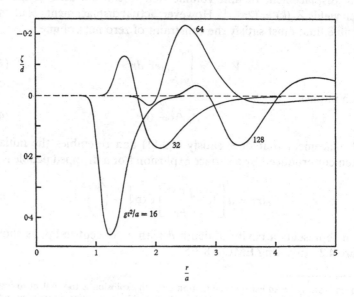

FIGURE 4.3   The asymptotic displacement described by (4.3.16) [from *J.
Fluid Mech. 34*, 368 (1968) by courtesy of Cambridge University Press].

Substituting (4.3.15) into (4.3.10), we obtain

$$\zeta(r, t) \sim 2^{-17/2} da^4 g^3 r^{-7} t^6 \exp\left[-\left(\frac{gat^2}{8r^2}\right)^2\right] \cos\frac{gt^2}{4r}$$

$$(\tfrac{1}{4}gt^2 \gg r \gg a). \quad (4.3.16)$$

This displacement is plotted in Figure 4.3.

EXERCISES

4.1 Replace $J_0(kr)$ in (4.2.10) by the asymptotic approximation

$$J_0(kr) \sim \left(\frac{2}{\pi kr}\right)^{1/2} \cos(kr - \tfrac{1}{4}\pi)$$

and use the method of stationary phase to obtain the asymptotic approximation

$$\phi \sim -iAac(R \sin \theta)^{-1} J_1\left(\frac{\omega a}{c} \sin \theta\right) \exp\left[i\omega\left(t - \frac{R}{c}\right)\right],$$

where R and θ ($\theta = 0$ on axis) are spherical polar coordinates. Calculate the radiated power by integrating the mean product of pressure and radial velocity over a hemisphere of radius $R \to \infty$ and show that the result agrees with (4.2.13).

*4.2 The piston in Section 4.2 is given the displacement $AH(t)$. Show that the resulting force on the piston is

$$F = -\rho c^2 A \left[4 - \left(\frac{ct}{a}\right)^2\right]^{1/2} H(2a - ct).$$

Note:

$$\int_0^\infty J_1^2(x) J_1(\mu x)\, dx = \tfrac{1}{2}\pi^{-1}(4 - \mu^2)^{1/2} H(2 - \mu).$$

4.3 If $V(s)$ and $F(s)$ are the Laplace transforms of the velocity of, and force on, the piston in Section 4.2, then

$$Z(s) = \frac{F(s)}{V(s)} \quad \text{and} \quad Y(s) = \frac{V(s)}{F(s)}$$

are the Laplace transforms of the *impulsive impedance* and *impulsive admittance* of the piston. Use the result of Exercise 4.2 to show that

$$Z(s) = 1 - \frac{4}{\pi} \int_0^1 e^{-2su}(1 - u^2)^{1/2}\, du$$

after rendering all variables appropriately nondimensional. Use this result to show that

$$\lim_{t \to \infty} y(t) = \frac{3\pi}{8}.$$

4.4   The asymptotic approximation of (4.3.6) requires $kr \gg 1$, which leaves some question as to the validity of (4.3.7), wherein k goes to zero. Repeat the derivation of (4.3.10) by substituting the integral representation

$$J_0(kr) = \frac{2}{\pi} \int_0^{\pi/2} \cos(kr \cos \alpha) \, d\alpha$$

into (4.3.5) and carrying out stationary-phase approximations with respect to each of k and α in that order.

4.5   The impulse of heat Q is supplied uniformly over the circle $r < a$ to the semi-infinite solid $z > 0$ at $t = 0$, after which the boundary $z = 0$ is insulated (no heat is transferred across it). Show that the subsequent temperature of the boundary is given by

$$v = \frac{Q}{Ka} \left(\frac{\kappa}{\pi^3 t} \right)^{1/2} \int_0^\infty J_0(kr) J_1(ka) \exp(-k^2 \kappa t) \, dk,$$

where K and κ are the conductivity and diffusivity of the solid.

Hint: The rate at which heat is transferred across $r < a$ is $(Q/\pi a^2)\delta(t)$.

5 FINITE FOURIER TRANSFORMS

5.1 Introduction

The transforms considered thus far are applicable to semi-infinite or infinite domains and have a common antecedent in Fourier's integral formula (1.2.6). It is natural to inquire whether transforms can be defined by (1.1.1) and their inverses derived from the theory of Fourier series. The essential result of this theory is that if the infinite sequence $\psi_1(x)$, $\psi_2(x)$, ... constitutes a complete orthogonal set of functions for the interval $a < x < b$ and the weighting function $w(x)$, corresponding to a discrete set of eigenvalues $p_1, p_2, \ldots$—that is, if

$$\int_a^b \psi_m(x)\psi_n(x)w(x)\,dx = \delta_{nm}N(p_n) \qquad (m, n = 1, 2, \ldots, \infty), \quad (5.1.1)$$

where δ_{nm} is the Kronecker delta, defined by

$$\delta_{nm} = 0, \quad m \neq n, \qquad \delta_{nm} = 1, \quad m = n, \qquad (5.1.2)$$

and

$$N(p_n) = \int_a^b \psi_n^2(x)w(x)\,dx, \qquad (5.1.3)$$

then it may be shown that

$$F(p) = \int_a^b f(x)\psi(x, p)w(x)\,dx \qquad (5.1.4a)$$

and

$$f(x) = \sum_{p_n} \frac{F(p_n)}{N(p_n)}\,\psi_n(x), \qquad \text{where} \qquad \psi_n(x) \equiv \psi(x, p_n), \quad (5.1.4b)$$

corresponding to $K(p, x) = \psi(x, p)w(x)$ in (1.1.1). Equations (5.1.4a, b) constitute a generalized, finite Fourier-transform pair.

The choice of the orthogonal functions $\psi(x, p)$ depends both on the differential equation and on the boundary conditions to be satisfied by $f(x)$, just as with the infinite transforms of the preceding chapters. We consider here finite sine and cosine transforms, which are appropriate to differential equations containing only even derivatives with respect to x and for which $w = 1$, and finite Hankel transforms, which are appropriate to the differential operator Δ_n of (4.1.2) and for which $w = r$, corresponding to that in the element of area $r\,dr\,d\theta$ in polar coordinates. The technique is generally applicable to all functions and boundary conditions of the Sturm–Liouville type and serves to mechanize much of the time-consuming detail associated with the determination of the unknown coefficients in the classical procedure that begins with separation of variables. The properties of several such transforms are tabulated in Table 2.4 (Appendix 2, p. 86).

5.2 Finite cosine and sine transforms

The simplest finite cosine and sine transforms are those for which $a = 0, b = \pi$, and $p = n$ in (5.1.4a, b), which then reduce to

$$F(n) = \int_0^\pi f(x)\begin{Bmatrix}\cos\\\sin\end{Bmatrix} nx\,dx \qquad (5.2.1a)$$

and

$$f(x) = \frac{1}{\pi}\sum_{n=0}^\infty (2 - \delta_{n0})F(n)\begin{Bmatrix}\cos\\\sin\end{Bmatrix} nx, \qquad (5.2.1b)$$

where either upper or lower alternatives must be taken together. The corresponding transforms of $f''(x)$ are given by

$$\int_0^\pi f''(x)\cos nx\,dx = -n^2 F(n) + (-)^n f'(\pi) - f'(0) \qquad (5.2.2)$$

and

$$\int_0^\pi f''(x)\sin nx\,dx = -n^2 F(n) + n[f(0) - (-)^n f(\pi)], \qquad (5.2.3)$$

whence these transforms are expedient for problems in which $f'(x)$ or $f(x)$, respectively, is prescribed at the end points of the interval. The generalization to an interval of length l merely requires a scale transformation, with x replaced by $\pi x/l$ (as in the following paragraph).

More general forms of these transforms, corresponding to Fourier's own generalization of his series, are given by

$$F(k) = \int_0^l f(x) \begin{cases} \cos \\ \sin \end{cases} kx \, dx, \qquad (5.2.4a)$$

and

$$f(x) = \sum_k (2 - \delta_{k0})(k^2 + h^2)[h + (k^2 + h^2)l]^{-1} F(k) \begin{cases} \cos \\ \sin \end{cases} kx, \quad (5.2.4b)$$

where

$$k \begin{cases} \tan \\ \cot \end{cases} kl = \pm h, \qquad (5.2.5)$$

$$\int_0^l f''(x) \begin{cases} \cos \\ \sin \end{cases} kx \, dx = -k^2 F(k) + [f'(l) + hf(l)] \begin{cases} \cos \\ \sin \end{cases} kl \begin{matrix} -f'(0) \\ +kf(0) \end{matrix}, \quad (5.2.6)$$

and either upper or lower alternatives must be taken together. These transforms are applicable to problems in heat conduction in which radiation takes place at $x = l$ and to problems involving lumped parameters at the boundaries of electrical or mechanical systems, such that $f'(l) + hf(l)$ is prescribed.

We remark that h usually is nonnegative in physical problems, by virtue of which (5.2.5) has only real roots, which occur in pairs of equal magnitude and opposite sign, with only the positive values being included in the summation of (5.2.4b). But if h is negative, (5.2.5) has a pair of conjugate imaginary roots, say $\pm i\kappa$, to which the corresponding term in (5.2.4b) is

$$f_\kappa(x) = \pm 2(\kappa^2 - h^2)[(\kappa^2 - h^2)l - h]^{-1} \begin{cases} \cosh \\ -\sinh \end{cases} \kappa x \int_0^l f(\xi) \begin{cases} \cosh \\ \sinh \end{cases} \kappa\xi \, d\xi.$$

$$(5.2.7)$$

We note, however, that $f(x)$ depends only on k^2 and therefore remains real. If $h = 0$, then the value $k = 0$ appears as a nontrivial root of (5.2.5) for the cosine transform, and (5.2.4b) must be replaced by (5.2.1b).

The roots of (5.2.5) are given approximately by

$$k \doteq \begin{cases} n\pi & + h(n\pi l)^{-1} \\ (n - \tfrac{1}{2})\pi + h[(n - \tfrac{1}{2})\pi l]^{-1} \end{cases} \qquad (n = 1, 2, \ldots) \qquad (5.2.8)$$

with an accuracy of 10% (1%) for all n ($n \geq 2$) if $hl \leq 1$ (the student should derive this result). These roots also are tabulated numerically [Abramowitz and Stegun, Table 4.20].

5.3 Wave propagation in a bar

We apply the finite Fourier transform to the problem of Section 2.8, as posed by

$$c^2 y_{xx} = y_{tt}, \tag{5.3.1}$$

$$y = 0 \quad \text{at } x = 0, \tag{5.3.2a}$$

$$E y_x = P \quad \text{at } x = l, \tag{5.3.2b}$$

and

$$y = y_t = 0 \quad \text{at } t = 0 \text{ and } 0 < x < l. \tag{5.3.3}$$

Referring to (5.2.4)–(5.2.6), with $f(x)$ replaced by $y(x, t)$, we find that the boundary conditions of (5.3.2) may be accommodated by using a finite sine transform with $h = 0$ and kl equal to an odd multiple of $\frac{1}{2}\pi$:

$$Y(k, t) = \int_0^l y(x, t) \sin kx \, dx \tag{5.3.4}$$

and

$$k = k_n = \frac{(2n + 1)\pi}{2l} \quad (n = 0, 1, \ldots). \tag{5.3.5}$$

Transforming (5.3.1)–(5.3.3), we obtain

$$Y_{tt} + (kc)^2 Y = \frac{Pc^2}{E} \sin kl \tag{5.3.6}$$

and

$$Y = Y_t = 0 \quad \text{at} \quad t = 0. \tag{5.3.7}$$

A particular solution of (5.3.6) is given by $(P/Ek^2) \sin kl$, so that the general solution is

$$Y = \frac{P}{Ek^2} \sin kl + A \cos kct + B \sin kct.$$

Invoking (5.3.7) to determine A and B, we obtain

$$Y(k, t) = \frac{P}{k^2 E} [1 - \cos(kct)] \sin kl. \tag{5.3.8}$$

Substituting $F = Y$ and kl from (5.3.5) and (5.3.8) into (5.2.4b) and setting $h = 0$, we obtain

$$y(x, t) = \frac{8Pl}{\pi^2 E} \sum_{n=0}^{\infty} (-)^n (2n + 1)^{-2} \sin k_n x [1 - \cos(k_n ct)]. \quad (5.3.9)$$

The terms in the series that are not time-dependent may be identified as the Fourier-series representation of Px/E, whence (5.3.9) may be reduced to (2.8.9).

5.4 Heat conduction in a slab

We illustrate the generalized transform described by (5.2.4)–(5.2.6) with $h \neq 0$ by considering one-dimensional conduction of heat in a slab of thickness l and initial temperature v_0, one face of which, $x = 0$, is insulated, and the other face of which, $x = l$, radiates into a medium at constant temperature, say $v = 0$ (the temperature in the slab being measured relative to that in $x > l$). The heat-conduction equation is, from (2.9.1),

$$v_t = \kappa v_{xx}, \quad (5.4.1)$$

where κ is the diffusivity. Invoking the condition of zero heat flow at $x = 0$ and Newton's law of cooling (which assumes that the relative temperature v is small compared with the absolute temperature), we obtain the boundary conditions

$$v_x = 0 \quad \text{at } x = 0 \quad (5.4.2a)$$

and

$$v_x = -hv \quad \text{at } x = l \quad (5.4.2b)$$

and the initial condition

$$v = v_0 \quad \text{at } t = 0. \quad (5.4.3)$$

Referring to (5.2.4)–(5.2.6), we find that the boundary conditions of (5.4.2) may be accommodated by using a finite cosine transform,

$$V(k, t) = \int_0^l v(x, t) \cos kx \, dx, \quad (5.4.4)$$

with

$$k \tan kl = h. \quad (5.4.5)$$

Transforming (5.4.1)–(5.4.3) with the aid of (5.2.6), we obtain

$$V_t + \kappa k^2 V = 0 \quad (5.4.6)$$

and

$$V = v_0 \int_0^l \cos kx \, dx = v_0 k^{-1} \sin kl \qquad \text{at } t = 0. \qquad (5.4.7)$$

Determining the coefficient of the exponential solution of (5.4.6) to satisfy (5.4.7), we obtain

$$V = v_0 k^{-1} \exp(-\kappa k^2 t) \sin kl. \qquad (5.4.8)$$

Inverting (5.4.8) with the aid of (5.2.4b), we obtain

$$v(x, t) = 2v_0 \sum_k \frac{(k^2 + h^2) \sin kl}{k[h + (k^2 + h^2)l]} \cos kx \exp(-\kappa k^2 t), \qquad (5.4.9)$$

where the summations are over the positive roots of (5.4.5).

5.5 Finite Hankel transforms

The Bessel functions $J_\nu(kr)$, $\nu \geq -\frac{1}{2}$, form a complete, orthogonal set for the interval $0 < r < a$ with the weighting function $w = r$ if the sequence $k_1, k_2, \ldots$ is determined by

$$J_\nu(ka) = 0 \qquad (0 < k_1 < k_2 < \ldots). \qquad (5.5.1)$$

The corresponding, series representation of the function $f(r)$ is known as a Fourier-Bessel series. Setting $x = r$, $\psi_n = J_\nu(kr)$, and $w = r$ in (5.1.3) and (5.1.4), we obtain the finite Hankel transform

$$F_\nu(k) = \int_0^a f(r) J_\nu(kr) r \, dr \qquad (5.5.2a)$$

and its inverse

$$f(r) = \frac{2}{a^2} \sum_k \frac{F_\nu(k) J_\nu(kr)}{J_{\nu-1}^2(ka)}. \qquad (5.5.2b)$$

The evaluation of N in passing from (5.1.4b) to (5.5.2b), and also (5.5.6) below, follows from the known integral

$$\int_0^x J_\nu^2(\xi) \xi \, d\xi = \frac{1}{2}(x^2 - \nu^2) J_\nu^3(x) + \frac{1}{2} x^2 J_\nu'^2(x). \qquad (5.5.3)$$

We also remark that (5.5.1) implies $J_\nu' = J_{\nu-1} = -J_{\nu+1}$ through the recursion formulas for J_ν and its derivative, J_ν'.

The finite Hankel transform, like its infinite counterpart of Section 4.1, arises naturally in connection with the Laplacian operator in cylindrical polar coordinates. Proceeding as in Section 4.1 and invoking (5.5.1) and $J'_\nu = J_{\nu-1}$ in the partially integrated terms, we obtain

$$\int_0^a (\Delta_\nu f) J_\nu(kr) r \, dr = -k^2 F_\nu(k) - ka J_{\nu-1}(ka) f(a), \qquad (5.5.4)$$

where the operator Δ_ν is defined by (4.1.2). We may replace $J_{\nu-1}$ by $-J_{\nu+1}$; this is convenient for $\nu = 0$.

A more general form of the finite Hankel transform, corresponding to the Dini-series representation of $f(r)$ in $0 < r < a$ [Watson (1945), Chapter 18] and analogous to the transforms defined by (5.2.4)–(5.2.6), is determined by the roots of

$$k J'_\nu(ka) + h J_\nu(ka) = 0, \qquad (5.5.5)$$

which leads to

$$f(r) = 2 \sum_k \frac{k^2 F_\nu(k) J_\nu(kr)}{[(k^2 + h^2)a^2 - \nu^2] J_\nu^2(ka)} \qquad (5.5.6)$$

in place of (5.5.2b).

The summation in (5.5.6) is over the positive roots of (5.5.5) if $h > 0$. If $h < 0$, (5.5.5) has a pair of conjugate imaginary roots, say $k = \pm i\kappa$, for which the corresponding (single) term in (5.5.6) is

$$f_\kappa = 2 \left[\frac{\kappa^2}{(\kappa^2 - h^2)a^2 + \nu^2} \right] \frac{I_\nu(\kappa r)}{I_\nu^2(\kappa a)} \int_0^a f(\eta) I_\nu(\kappa \eta) \eta \, d\eta, \qquad (5.5.7)$$

where I_ν is a modified Bessel function. The counterpart of (5.5.4) is

$$\int_0^a (\Delta_\nu f) J_\nu(kr) r \, dr = a J_\nu(ka)(f_r + hf)_{r=a} - k^2 F_\nu(k) \qquad (5.5.8a)$$

$$= -ka J'_\nu(ka)(h^{-1} f_r + f)_{r=a} - k^2 F_\nu(k), \qquad (5.5.8b)$$

where (5.5.8a) and (5.5.8b) are equivalent by virtue of (5.5.5).

5.6 Cooling of a circular bar

We now consider the axisymmetric, and physically more interesting, counterpart of the heat-conduction problem of Section 5.4. An infinitely long, circular cylinder of radius a, diffusivity κ, and initial temperature v_0 radiates into a medium at zero (relative) temperature according to Newton's law of cooling. Invoking axial symmetry in the three-dimensional generalization of (2.9.1) [see footnote following (2.9.1)], we obtain the heat-conduction equation

$$v_t = \kappa \Delta_0 v \equiv \kappa(v_{rr} + r^{-1}v_r), \tag{5.6.1}$$

the boundary condition

$$v_r = -hv \qquad \text{at } r = a \text{ and } t > 0, \tag{5.6.2}$$

and the initial condition

$$v = v_0 \qquad \text{at } t = 0 \text{ and } 0 < r < a. \tag{5.6.3}$$

The form of (5.6.1) and (5.6.2) suggests the finite Hankel transform

$$V(k, t) = \int_0^a v(r, t)J_0(kr)r \, dr, \tag{5.6.4}$$

where the k are determined by (5.5.5) with $v = 0$ or, equivalently,

$$kJ_1(ka) = hJ_0(ka). \tag{5.6.5}$$

Transforming (5.6.1)-(5.6.3) with the aid of (5.5.8a), we obtain

$$V_t + \kappa k^2 V = 0 \tag{5.6.6}$$

and

$$V = v_0 \int_0^a J_0(kr)r \, dr = v_0 k^{-1}aJ_1(ka) \qquad \text{at } t = 0, \tag{5.6.7}$$

the solution of which is [cf. the solution of (5.4.6) and (5.4.7)]

$$V = v_0 k^{-1}aJ_1(ka)\exp(-\kappa k^2 t). \tag{5.6.8}$$

Setting $v = 0$, replacing F_v by V in (5.5.6), and eliminating J_1 through (5.6.5), we obtain

$$v(r, t) = \frac{2v_0 h}{a} \sum_k \frac{J_0(kr)\exp(-\kappa k^2 t)}{(k^2 + h^2)J_0(ka)}, \tag{5.6.9}$$

where the summation is over the positive roots of (5.6.5).

5.7 Viscous diffusion in a rotating cylinder †

The axisymmetric motion of a viscous fluid in concentric circles about the axis of rotation of an infinitely long cylinder is governed by [Lamb (1932), Section 328a(5)]

$$v_t = \nu \Delta_1 v \equiv \nu(v_{rr} + r^{-1}v_r - r^{-2}v), \tag{5.7.1}$$

where v is the tangential velocity, and ν is the kinematic viscosity (not to

---

† Sneddon (1951) gives other examples of this type.

be confused with the order of the Bessel function in Section 5.5). We consider the *spin-up* of a cylinder of fluid of radius a, which is initially at rest, following the imposition of the angular velocity Ω at the outer boundary. The corresponding boundary and initial conditions are

$$v = \Omega a \quad \text{at } r = a \text{ and } t > 0 \tag{5.7.2}$$

and

$$v = 0 \quad \text{at } t = 0 \text{ and } 0 < r < a. \tag{5.7.3}$$

The form of (5.7.1) and (5.7.2) suggests the finite Hankel transform

$$V(k, t) = \int_0^a v(r, t) J_1(kr) r \, dr, \tag{5.7.4}$$

where the k are determined by the zeros of

$$J_1(ka) = 0. \tag{5.7.5}$$

Transforming (5.7.1)–(5.7.3) with the aid of (5.5.4), we obtain

$$V_t + vk^2 V = -v\Omega a^2 k J_0(ka) \tag{5.7.6}$$

and

$$V = 0 \quad \text{at } t = 0. \tag{5.7.7}$$

Remarking that a particular solution of (5.7.6) is given by $-\Omega a^2 k^{-1} J_0(ka)$ and choosing the complementary solution to satisfy (5.7.7), we obtain

$$V(k, t) = -\Omega a^2 k^{-1} J_0(ka)[1 - \exp(-vk^2 t)]. \tag{5.7.8}$$

Setting $v = 1$ in (5.5.2b), we obtain

$$v(r, t) = -2\Omega \sum_k [k J_0(ka)]^{-1} J_1(kr)[1 - \exp(-vk^2 t)], \tag{5.7.9}$$

where the summation is over the positive roots of (5.7.5).

We remark that r has the corresponding representation

$$r = -2 \sum_k [k J_0(ka)]^{-1} J_1(kr), \tag{5.7.10}$$

so that $v \sim \Omega r$ as $t \to \infty$, as also may be inferred from physical considerations.

5.8 Conclusion

The superiority of the finite-transform method, over either the classical procedure of separation of variables or the Laplace transformation, in obtaining solutions as expansions of natural modes tends to increase with the complexity of the problem. In particular, the finite-transform method always provides the modal expansion of the static solution, as in (5.3.9) above; however, this may not always be an advantage. We emphasize, nevertheless, that the Laplace transform is both more flexible and more powerful. It both incorporates alternative interpretations, such as the traveling-wave expansion of (2.8.13), and places less stringent conditions on the boundary conditions that may be accommodated.

EXERCISES

5.1   A uniform string of length l and line density ρ is stretched between two fixed points, $x = 0$ and $x = l$, to tension ρc^2. It is displaced a small distance a at a point distant b from the origin and released at $t = 0$. Given the equation of motion,

$$c^2 y_{xx} = y_{tt},$$

show that the subsequent displacement is

$$y(x, t) = \frac{2al^2}{\pi^2 b(l - b)} \sum_{n=1}^{\infty} n^{-2} \sin \frac{n\pi b}{l} \sin \frac{n\pi x}{l} \cos \frac{n\pi ct}{l}.$$

5.2   A string of line density ρ and length l is stretched to tension ρc^2. The end $x = 0$ is fixed, and the end $x = l$ is attached to a massless ring that is free to slide on a smooth rod. At $t = 0$, when the system is at rest with the ring displaced a small distance a from equilibrium position, the ring is released. Show that the subsequent displacement of the string is

$$y(x, t) = \frac{8a}{\pi^2} \sum_{n=0}^{\infty} (-)^n (2n + 1)^{-2} \sin \frac{(2n + 1)\pi x}{2l} \cos \frac{(2n + 1)\pi ct}{2l}.$$

5.3   A string of line density ρ and length l is stretched to tension ρc^2. The end $x = l$ is fixed and at $t = 0$, when the string is at rest in its equilibrium position, the end $x = 0$ is given a small oscillation $a \sin \omega t$. Show that the subsequent displacement of the point x is

$$\frac{a \sin \omega t \sin[\omega(l - x)/c]}{\sin(\omega l/c)} + \sum_{r=1}^{\infty} \frac{2lca\omega}{\omega^2 l^2 - \pi^2 r^2 c^2} \sin \frac{r\pi x}{l} \sin \frac{r\pi ct}{l}.$$

5.4   (a) Show that (5.4.9) has the alternative representation

$$v = 2v_0 h \sum_k [h + (k^2 + h^2)l]^{-1} \sec kl \cos kx \exp(-\kappa k^2 t)$$

by solving (5.4.5) for $\sin kl$ and $\cos kl$. (b) Show that (5.4.9) reduces to

$$v = \frac{4v_0}{\pi} \sum_{n=0}^{\infty} (-)^n (2n + 1)^{-1} \cos k_n x \exp(-\kappa k_n^2 t),$$

where $k_n = (2n + 1)(\pi/2l)$, for $h = \infty$.

5.5 Determine the solution to the diffusion equation

$$v_t = \kappa v_{xx}$$

with

$$v = \begin{cases} x & \text{for } 0 < x < \tfrac{1}{2}l \text{ at } t = 0 \\ l - x & \text{for } \tfrac{1}{2}l < x < l \text{ at } t = 0 \end{cases}$$

and

$$v = 0 \qquad \text{at } x = 0 \text{ and } l \text{ and } t > 0$$

by (a) Laplace transformation and (b) finite-Fourier transformation.
Answer:

$$v = \frac{4l}{\pi^2} \sum_{n=0}^{\infty} \frac{(-)^n}{(2n+1)^2} \exp\left[-\frac{(2n+1)^2\pi^2\kappa t}{l^2}\right] \sin\frac{(2n+1)\pi x}{l}.$$

5.6 Consider the problem of Section 5.4 with the boundary condition $v = f(t)$ in place of (5.4.2a) and the initial condition $v = 0$ in place of (5.4.3). Show that

$$v(x, t) = 2\kappa \sum_k k(k^2 + h^2)[h + (k^2 + h^2)l]^{-1} \sin kx \int_0^t \exp[-\kappa k^2(t - \tau)]f(\tau)\,d\tau,$$

where

$$k \cot kl = -h.$$

5.7 Solve Exercise 2.17 by using a finite Fourier transform.
5.8 Solve Exercise 2.20 by using a finite Fourier transform.
5.9 Use the Laplace transform to solve the problem of Section 5.6.
5.10 The top $(0 < x \leq l)$ and bottom $(-l \leq x < 0)$ portions of a cylindrical box bounded by $x = \pm l$ and $r = a$ are insulated from one another and charged to the potentials v_0 and $-v_0$, respectively. Show that the potential inside the box, which must satisfy Laplace's equation,

$$v_{xx} + v_{rr} + r^{-1}v_r = 0,$$

is given by either

$$v = \frac{2v_0}{a}\,\mathrm{sgn}\,x \sum_k \left[1 - \frac{\sinh k(l - |x|)}{\sinh kl}\right]\frac{J_0(kr)}{kJ_1(ka)},$$

where the summation is over the positive roots of $J_0(ka) = 0$, or

$$v = \frac{2v_0}{\pi} \sum_{n=1}^{\infty} n^{-1}\left[\frac{I_0(n\pi r/l)}{I_0(n\pi a/l)} + (-)^{n-1}\right] \sin\frac{n\pi x}{l},$$

where I_0 is a modified Bessel function.

5.11   The functions

$$Z_n(kr) = J_n(kr)Y_n(ka) - Y_n(kr)J_n(ka)$$

form a complete orthogonal set in $a < r < b$ if the k are given by the positive roots of

$$Z_n(kb) = 0.$$

Obtain the corresponding finite-transform pair (see Table 2.4). Use the result to determine the temperature distribution in a long annular tube of diffusivity κ if the inner and outer surfaces, $r = a$ and $r = b$, are held at zero temperature and the initial temperature is v_0.

Answer:

$$v = \pi v_0 \sum_k \left[J_0(ka) + J_0(kb) \right]^{-1} J_0(kb) Z_0(kr) \exp(-\kappa k^2 t).$$

5.12   Suppose that the viscous fluid of Section 5.7 is bounded internally by a stationary cylinder of radius a and externally by a cylinder of radius b that begins to rotate with the angular velocity Ω at $t = 0$. Show that

$$v(r, t) = \pi \Omega b \sum_k \frac{J_1(ka)J_1(kb)Z_1(kr)[1 - \exp(-vk^2 t)]}{[J_1^2(ka) - J_1^2(kb)]},$$

where

$$Z_1(kr) = J_1(kr)Y_1(ka) - Y_1(kr)J_1(ka),$$

and the k are determined by the positive zeros of $Z_1(kb)$.

PARTIAL-FRACTION EXPANSIONS
APPENDIX 1

Let

$$F(p) = \frac{G(p)}{H(p)}, \tag{A1.1}$$

where

$$G(p) = g_0 + g_1 p + \cdots + g_M p^M \tag{A1.2}$$

and

$$H(p) = h_0 + h_1 p + \cdots + h_N p^N \qquad (N > M) \tag{A1.3}$$

are polynomials with no common factors (common factors may be cancelled if originally present), and the degree of H is higher than that of G $(N > M)$.

We consider first the case where the zeros of $H(p)$, say $p_1, p_2, \ldots, p_N$, are all distinct. It then follows from known results in algebra that $H(p)$ can be factored to obtain

$$H(p) = h_N(p - p_1)(p - p_2) \cdots (p - p_N) \tag{A1.4}$$

and that $F(p)$ can be developed in a partial-fraction expansion of the form

$$F(p) = \frac{C_1}{p - p_1} + \frac{C_2}{p - p_2} + \cdots + \frac{C_N}{p - p_N}. \tag{A1.5}$$

The C_n, $n = 1, 2, \ldots, N$, in the expansion (A1.5) may be determined by one of the following procedures: (a) multiplying both sides of (A1.5) through by $H(p)$, equating the coefficients of p^n, $n = 1, 2, \ldots, M$ on both sides of the resulting equation, and requiring the coefficients of p^n, $n = M + 1, \ldots, N$ on the right-hand side thereof to vanish identically; (b) multiplying both sides of (A1.5) through by $p - p_n$ and letting $p \to p_n$ to obtain

$$C_n = \lim_{p \to p_n} \frac{(p - p_n)G(p)}{H(p)} \equiv \lim_{p \to p_n}(p - p_n)F(p) \qquad (n = 1, 2, \ldots, N); \quad \text{(A1.6)}$$

(c) invoking the identity (in effect, L'Hospital's rule)

$$\lim_{p \to p_n} \frac{H(p) - H(p_n)}{p - p_n} = H'(p)|_{p = p_n} \equiv H'(p_n) \qquad \text{(A1.7)}$$

in (A1.6) to obtain

$$C_n = \frac{G(p_n)}{H'(p_n)}. \qquad \text{(A1.8)}$$

The procedure (a) is efficient only in relatively simple cases, although it occasionally may be more direct than either (b) or (c). The latter procedures are almost equivalent, but (c) is generally more efficient.

Substituting (A1.8) into (A1.5), we obtain the general result

$$F(p) = \sum_{n=1}^{N} \frac{G(p_n)}{H'(p_n)(p - p_n)} \qquad [H(p_n) = 0]. \qquad \text{(A1.9)}$$

The inverse transform of this expansion is given by (2.7.7), which, is, however, of greater generality [see discussion preceding (2.7.7)].

Consider, for example, the transform of (2.3.7),

$$F(p) = \frac{1}{p(p^2 + \beta^2)}. \qquad \text{(A1.10)}$$

Choosing $G(p) = 1$ and

$$H(p) = p(p^2 + \beta^2) = p(p - i\beta)(p + i\beta), \qquad \text{(A1.11)}$$

we have $M = 0$, $N = 3$, $p_1 = 0$, $p_2 = i\beta$, $p_3 = -i\beta$, and

$$F(p) = \frac{1}{H(p)} = \frac{C_1}{p} + \frac{C_2}{p - i\beta} + \frac{C_3}{p + i\beta}. \qquad \text{(A1.12)}$$

Considering first procedure (a), we multiply both sides of (A1.12) through by $H(p)$ in its factored form to obtain the identity

$$1 = C_1(p - i\beta)(p + i\beta) + C_2 p(p + i\beta) + C_3 p(p - i\beta)$$

$$= C_1\beta^2 + (C_2 - C_3)i\beta p + (C_1 + C_2 + C_3)p^2. \qquad \text{(A1.13)}$$

Equating the coefficients of p^0 and requiring the coefficients of p and p^2 on the right-hand side of (A1.13) to vanish, we obtain

$$1 = C_1\beta^2, \qquad C_2 - C_3 = 0, \qquad \text{and} \qquad C_1 + C_2 + C_3 = 0, \quad \text{(A1.14)}$$

the solution of which yields

$$C_1 = \frac{1}{\beta^2} \quad \text{and} \quad C_2 = C_3 = -\frac{1}{2\beta^2}. \tag{A1.15}$$

Substituting (A1.15) into (A1.12), we obtain

$$F(p) = \frac{1}{\beta^2}\left[\frac{1}{p} - \frac{1}{2(p-i\beta)} - \frac{1}{2(p+i\beta)}\right] \tag{A1.16a}$$

$$= \frac{1}{\beta^2}\left(\frac{1}{p} - \frac{p}{p^2+\beta^2}\right). \tag{A1.16b}$$

the latter form being that of (2.3.7). Considering procedure (b), we substitute $G = 1$ and $H = p(p^2 + \beta^2)$ in (A1.6) to obtain

$$C_1 = \lim_{p\to 0}\frac{(p-0)}{p(p^2+\beta^2)} = \frac{1}{\beta^2}.$$

$$C_2 = \lim_{p\to i\beta}\frac{(p-i\beta)}{p(p^2+\beta^2)} = \lim_{p\to i\beta}\frac{1}{p(p+i\beta)} = -\frac{1}{2\beta^2}.$$

and

$$C_3 = \lim_{p\to -i\beta}\frac{(p+i\beta)}{p(p^2+\beta^2)} = \lim_{p\to -i\beta}\frac{1}{p(p-i\beta)} = -\frac{1}{2\beta^2}. \tag{A1.17}$$

Considering procedure (c), we substitute $G = 1$ and $H' = 3p^2 + \beta^2$ in (A1.8) to obtain

$$C_1 = \frac{1}{\beta^2}, \quad C_2 = \frac{1}{3(i\beta)^2 + \beta^2} = -\frac{1}{2\beta^2}.$$

and

$$C_3 = \frac{1}{3(-i\beta)^2 + \beta^2} = -\frac{1}{2\beta^2}. \tag{A1.18}$$

It is evident, even in this rather simple example, that (a) is the most cumbersome procedure, whereas (c) is the most efficient.

Various special cases, and generalizations, of (A1.9) are given by Gardner–Barnes (1942), pp. 153–63. The most important generalization allows for a double zero of $H(p)$, say $p_1 \equiv p_2$, such that (A1.4) is replaced by

$$H(p) = h_N(p-p_1)^2(p-p_3)\cdots(p-p_N). \tag{A1.19}$$

The expansion (A1.5) then must be replaced by

$$F(p) = \frac{C_{11}}{p-p_1} + \frac{C_{12}}{(p-p_1)^2} + \frac{C_3}{p-p_3} + \cdots + \frac{C_N}{p-p_N}. \tag{A1.20}$$

Procedure (a) above may be applied directly to this expansion to obtain

$C_{11}, C_{12}, C_3, \ldots, C_N$. Procedures (b) and (c) may be applied to obtain $C_3, \ldots, C_N$, but must be modified to obtain C_{11} and C_{12}. Multiplying both sides of (A1.20) through by $(p - p_1)^2$ and introducing the function

$$\Phi(p) = (p - p_1)^2 F(p) \equiv \frac{(p - p_1)^2 G(p)}{H(p)}, \qquad \text{(A1.21)}$$

we obtain

$$\Phi(p) = C_{12} + C_{11}(p - p_1) + (p - p_1)^2 \left[\frac{C_3}{p - p_3} + \cdots + \frac{C_N}{p - p_N} \right].$$

$$\text{(A1.22)}$$

Setting $p = p_1$ in (A1.22), we obtain

$$C_{12} = \Phi(p_1). \qquad \text{(A1.23)}$$

Differentiating (A1.22) with respect to p and then setting $p = p_1$, we obtain

$$C_{11} = \Phi'(p_1). \qquad \text{(A1.24)}$$

Substituting (A1.23) and (A1.24), together with (A1.8) for $n = 3, \ldots, N$, into (A1.20), we obtain

$$F(p) = \frac{\Phi(p_1)}{(p - p_1)^2} + \frac{\Phi'(p_1)}{p - p_1} + \sum_{n=3}^{N} \frac{G(p_n)}{H'(p_n)(p - p_n)}, \qquad \text{(A1.25)}$$

where $\Phi(p)$ is given by (A1.21). See also Exercise 2.8.

TABLES
APPENDIX 2

Table 2.1 Laplace-transform pairs

(Greek letters in the following may be complex as noted.)

	$f(t)$	$F(p) = \int_0^\infty e^{-pt} f(t)\, dt$
2.1.1	$t^\nu\ (\mathscr{R}\nu > -1)$	$\Gamma(\nu + 1)p^{-\nu - 1}$
2.1.2	$e^{-\alpha t}(\mathscr{R}\alpha \geqq 0)$	$(p + \alpha)^{-1}$
2.1.3	$\cos bt$	$p(p^2 + b^2)^{-1}$
2.1.4	$\sin bt$	$b(p^2 + b^2)^{-1}$
2.1.5	$e^{-at}\cos bt$	$(p + a)[(p + a)^2 + b^2]^{-1}$
2.1.6	$e^{-at}\sin bt$	$b[(p + a)^2 + b^2]^{-1}$
2.1.7	$te^{-\alpha t}\ (\mathscr{R}\alpha \geqq 0)$	$(p + \alpha)^{-2}$
2.1.8	$\delta(t)$	1
2.1.9	$t^m \log t$	$m!\,p^{-m-1}[1 + \frac{1}{2} + \frac{1}{3} + \cdots + m^{-1} - \gamma - \log p]$
2.1.10	$\operatorname{erfc}(\frac{1}{2}at^{-1/2})$	$p^{-1}\exp(-ap^{1/2})$
2.1.11	$J_0(at)$	$(p^2 + a^2)^{-1/2}$
2.1.12	$J_0[a(t^2 - b^2)^{1/2}]H(t - b)$	$(p^2 + a^2)^{-1/2}\exp[-b(p^2 + a^2)^{1/2}]$

Table 2.2 Operational formulas

	$f(t)$	$F(p) = \int_0^\infty e^{-pt} f(t)\, dt$
2.2.1	$\alpha f(t)$	$\alpha F(p)$
2.2.2	$f_1(t) + f_2(t)$	$F_1(p) + F_2(p)$
2.2.3	$f^{(n)}(t)$	$p^n F(p) - \sum_{m=0}^{n-1} p^{n-m-1} f^{(m)}(0)$
2.2.4	$\int_0^t f(\tau)\, d\tau$	$p^{-1} F(p)$
2.2.5	$f(t-a)H(t-a)\ (a \geqq 0)$	$e^{-ap} F(p)$
2.2.6	$e^{-\alpha t} f(t)$	$F(p+\alpha)$
2.2.7	$f(at)$	$a^{-1} F\left(\dfrac{p}{a}\right)$
2.2.8	$f(t+T) = f(t)$	$(1 - e^{-pT})^{-1} \int_0^T e^{-pt} f(t)\, dt$
2.2.9	$\int_0^t f_1(t-\tau) f_2(\tau)\, d\tau$	$F_1(p) F_2(p)$
2.2.10	$t^n f(t)$	$(-)^n F^{(n)}(p)$
2.2.11	$\dfrac{\partial}{\partial \alpha} f(t; \alpha)$	$\dfrac{\partial}{\partial \alpha} F(p; \alpha)$

Table 2.3 Infinite integral transforms

Transform	$F(p) = \mathcal{T}\{f(x)\}$	$f(x) = \mathcal{T}^{-1}\{F(p)\}$	Transforms of typical derivatives	Convolution theorem $\mathcal{T}^{-1}\{F(p)G(p)\}$
Laplace	$\displaystyle\int_0^\infty f(x)e^{-px}\,dx$	$\displaystyle\frac{1}{2\pi i}\int_{c-i\infty}^{c+i\infty} F(p)e^{px}\,dp$	$\mathcal{T}\{f^{(n)}\} = p^n F(p) - \sum_{m=0}^{n-1} p^m f^{(n-m-1)}(0)$	$\displaystyle\int_0^x f(\xi)g(x-\xi)\,d\xi$
Fourier	$\displaystyle\int_{-\infty}^\infty f(x)e^{-ipx}\,dx$	$\displaystyle\frac{1}{2\pi}\int_{-\infty}^\infty F(p)e^{ipx}\,dp$	$\mathcal{T}\{f^{(n)}\} = (ip)^n F(p)$	$\displaystyle\int_{-\infty}^\infty f(\xi)g(x-\xi)\,d\xi$
Fourier-cosine	$\displaystyle\int_0^\infty f(x)\cos px\,dx$	$\displaystyle\frac{2}{\pi}\int_0^\infty F(p)\cos px\,dp$	$\mathcal{T}\{f^{(2n)}\} = (-)^n p^{2n} F(p) - \sum_{m=0}^{n-1} (-)^m p^{2m} f^{(2n-2m-1)}(0)$	$\displaystyle\frac{1}{2}\int_0^\infty f(\xi)[g(|x-\xi|) + g(x+\xi)]\,d\xi$
Fourier-sine	$\displaystyle\int_0^\infty f(x)\sin px\,dx$	$\displaystyle\frac{2}{\pi}\int_0^\infty F(p)\sin px\,dp$	$\mathcal{T}\{f^{(2n)}\} = (-)^n p^{2n} F(p) - \sum_{m=1}^{n} (-)^m p^{2m-1} f^{(2n-2m)}(0)$	None available, but see Sneddon (1951), Section 3.6, for related theorems
Mellin	$\displaystyle\int_0^\infty f(x)x^{p-1}\,dx$	$\displaystyle\frac{1}{2\pi i}\int_{c-i\infty}^{c+i\infty} F(p)x^{-p}\,dp$	$\mathcal{T}\{x^n f^{(n)}\} = (-1)^n p(p+1)\cdots(p+n-1)F(p)$	$\displaystyle\int_0^\infty f(\xi)g\left(\frac{x}{\xi}\right)\frac{d\xi}{\xi}$
Hankel	$\displaystyle\int_0^\infty f(x)J_n(px)x\,dx$	$\displaystyle\int_0^\infty F(p)J_n(px)p\,dp$	$\mathcal{T}\left\{f'' + \frac{1}{x}f' - \left(\frac{n}{x}\right)^2 f\right\} = -p^2 F$	No simple result

Table 2.4 Finite integral transforms

Transform	$F(k) = \mathcal{T}\{f(x)\}$	$f(x) = \mathcal{T}^{-1}\{F(k)\}$	Eigenvalue equation for k_n ($k_n \geqq 0$)	Transforms of typical derivatives
Cosine	$\displaystyle\int_0^l f(x)\cos kx\,dx$	$\displaystyle\sum_{k_n}\frac{(2-\delta_{k0})(k^2+h^2)F(k)\cos kx}{h+l(k^2+h^2)}$	$k\tan kl = h$	$\mathcal{T}\{f''(x)\} = -k^2 F(k) - f'(0)$ $\qquad + [f'(l)+hf(l)]\cos kl$
Sine	$\displaystyle\int_0^l f(x)\sin kx\,dx$	$\displaystyle 2\sum_{k_n}\frac{(k^2+h^2)F(k)\sin kx}{h+l(k^2+h^2)}$	$k\cot kl = -h$	$\mathcal{T}\{f''(x)\} = -k^2 F(k) + kf(0)$ $\qquad + [f'(l)+hf(l)]\sin kl$
Hankel	$\displaystyle\int_0^a f(x)J_n(kx)x\,dx$	$\displaystyle 2\sum_{k_n}\frac{k^2 F(k)J_n(kx)}{[(h^2+k^2)a^2 - n^2]J_n^2(ka)}\;(h>0)$	$kJ_n'(ka) + hJ_n(ka) = 0$	$\mathcal{T}\left\{f'' + x^{-1}f' - \left(\dfrac{n}{x}\right)^2\right\} = -k^2 F(k)$ $\qquad + [f'(a)+hf(a)]aJ_n(ka)$
Annular Hankel	$\displaystyle\int_a^b f(x)Z_n(kx)x\,dx,$ $Z_n(kx) = Y_n(ka)J_n(kx)$ $\qquad - J_n(ka)Y_n(kx)$	$\displaystyle\frac{1}{2}\pi^2\sum_{k_n}\frac{k^2 J_n^2(kb)F(k)Z_n(kx)}{J_n^2(ka)-J_n^2(kb)}$	$Z_n(kb) = 0\;(b>a)$	$\mathcal{T}\left\{f'' + x^{-1}f' - \left(\dfrac{n}{x}\right)^2 f\right\} = -k^2 F(k)$ $\qquad + \dfrac{2}{\pi}\left[\dfrac{J_n(ka)}{J_n(kb)}f(b) - f(a)\right]$
Legendre	$\displaystyle\int_{-1}^1 f(x)P_n(x)\,dx$	$\displaystyle\frac{1}{2}\sum_{n=0}^\infty (2n+1)F(n)P_n(x)$	$k = n$	$\mathcal{T}\{[(1-x^2)f'(x)]'\} = -n(n+1)F(n)$

LIST OF NOTATIONS
APPENDIX 3

The numbers after each entry indicate either the equation or the section in which that entry is defined. Symbols used only in a local context, wherein they are defined explicitly, may not be included in this list.

C	closed contour of integration in complex plane
E	Young's modulus for elastic solid (Section 2.8)
EMOT	Erdélyi, Magnus, Oberhettinger, and Tricomi
F	typically an integral transform of $f(x)$
$\mathscr{F}, \mathscr{F}^{-1}$	Fourier-transform operator and its inverse (1.3.1)
$\mathscr{F}_c, \mathscr{F}_c^{-1}$	Fourier-cosine-transform operator and its inverse (1.3.5)
$\mathscr{F}_s, \mathscr{F}_s^{-1}$	Fourier-sine-transform operator and its inverse (1.3.6)
H	Heaviside step function (2.5.3)
$\mathscr{H}, \mathscr{H}^{-1}$	Hankel-transform operator and its inverse
I_n	modified Bessel function of first kind and nth order
J_n	Bessel function of first kind and nth order
K	transform kernel (1.1.1), thermal conductivity
$\mathscr{L}, \mathscr{L}^{-1}$	Laplace-transform operator and its inverse (1.4.3)
R	spherical radius; also residue
$\mathscr{R}$	operator that implies *real part* of its operand
T	entry in table (e.g., T2.1.1 implies entry 1 in Table 2.1); period of periodic function
$\mathscr{T}, \mathscr{T}^{-1}$	integral-transform operator and its inverse
Y_n	Bessel function of second kind and nth order
c	(1.4.1) and Figure 1.2; wave speed; specific heat (Section 2.9)
e	base of natural logarithms $= 2.7183\cdots$
erf z	$= 2\pi^{-1/2}\int_0^z \exp(-u^2)\,du = $ error function
erfc z	$= 1 - $ erf $z = $ complementary error function
f	typically an arbitrary function to be transformed (1.1.1)

$f^{(n)}(x)$	nth derivative of $f(x)$
$f^{(n)}(0)$	nth derivative of $f(x)$ evaluated at $x = 0$
f'	$\equiv f^{(1)}$
f''	$\equiv f^{(2)}$
g	gravitational acceleration
h	(5.2.5)
i	imaginary unit; subscript i denotes imaginary part of corresponding variable
k	Fourier-transform variable (1.3.1), (1.3.5), (1.3.6); spring constant in Section 2.3
k_n	discrete value of k in the monotonically increasing sequence $k_1, k_2, \cdots$
l	length
m, n	integers
p	transform parameter (1.1.1)
p_n	a pole of $F(p)$ or a discrete value of p in the monotonically increasing sequence $p_1, p_2, \cdots$
r	radius in polar coordinates; subscript r denotes real part of corresponding variable
s	index of summation; s takes only odd values ($s = 1, 3, \cdots$) in Section 2.6; also subscript denoting point of stationary phase
sgn	signum function (3.7.4)
x	independent variable; typically, but not necessarily spatial
x, y, z	Cartesian coordinates
Γ	gamma function; $\Gamma(n + 1) = n!$
Δ_n	(4.1.2)
Σ	summation sign
γ	Euler's constant $= 0.577215\cdots$
δ	Dirac delta function (2.4.2)
δ_{mn}	Kronecker delta (5.1.1)
ε	a small parameter that tends to zero (through positive values unless otherwise stated)
ζ	water-wave displacement (Section 3.6)
θ	angle, typically a polar coordinate
κ	diffusivity (2.9.1)
ρ	density (mass per unit volume or, for string, per unit length)
ω	angular frequency
∇	vector-gradient operator
∇^2	Laplacian operator; $\nabla^2 = (\partial/\partial x)^2 + (\partial/\partial y)^2 + (\partial/\partial z)^2$ in Cartesian coordinates; see (4.1.3) for polar coordinates
$\sim$	implies *is asymptotic to*
$\equiv$	implies *equal by definition*
$\doteq$	implies *approximately equal*

GLOSSARY

boundary condition   a condition imposed on a bounding surface (in three dimensions) or line (in two dimensions) or at a bounding point (in one dimension)

convolution theorem   see (2.4.1), (3.3.2), (3.3.7), and Table 2.3

finiteness condition   a degenerate form of a boundary condition, whereby a dependent variable is required to be finite in some limit

Heaviside's expansion theorem   see (2.7.7)

Heaviside's shifting theorem   see (2.5.2)

initial condition   a condition imposed at $t = 0$ or, more generally, at the origin of a *timelike* variable (as in Section 2.10)

Jordan's lemma   Let L denote the semicircle $|p| = P$ in $p_r < 0$; then

$$\lim_{P \to \infty} \int_L e^{ap} f(p)\, dp = 0 \qquad (a > 0)$$

provided that $f(Pe^{i\theta})$ vanishes uniformly in $\frac{1}{2}\pi \leq \theta \leq \frac{3}{2}\pi$ as $P \to \infty$.

L'Hospital's rule   If $f(a) = g(a) = 0$, $\lim_{x \to a} [f(x)/g(x)] = f'(a)/g'(a)$.

meromorphic   a function of a complex variable for which every point in the finite, complex plane is either a regular point (in the neighborhood of which the function is analytic) or a pole

radiation condition   a requirement that a disturbance appear as an outgoing wave; e.g. $f(r - ct)$ is an outgoing wave as $r \to \infty$, whereas $f(r + ct)$ is an incoming wave.

residue

$$R_* = \lim_{p \to p_*} \frac{d^{n-1}}{dp^{n-1}} \left[\frac{(p - p_*)^n F(p)}{(n-1)!} \right],$$

where p_* is a pole of nth order.

stationary phase   see Section 3.7

Watson's lemma   see (2.7.13)

BIBLIOGRAPHY

Texts and treatises

BECKENBACH, E. F. (Ed.), *Modern Mathematics for the Engineer,* Second Series (New York: McGraw-Hill, 1961).

BRACEWELL, RON, *The Fourier Transform and its Applications* (New York: McGraw-Hill, 1965).
A stimulating treatment from the viewpoint of modern electrical engineering.

CARSLAW, H. S., and J. C. JAEGER, *Operational Methods in Applied Mathematics* (New York: Oxford Univ. Press, 1953).
Deals only with the Laplace transform. Contains extensive collection of worked and unworked problems involving both ordinary and partial differential equations.

CARSLAW, H. S., and J. C. JAEGER, *Conduction of Heat in Solids* (New York: Oxford University Press, 1949).
Standard treatise on mathematical theory of heat conduction in solids. Extensive application of Laplace transform.

CHURCHILL, R. V., *Fourier Series and Boundary Value Problems* (New York: McGraw-Hill, 1963).

CHURCHILL, R. V., *Operational Mathematics* (New York: McGraw-Hill, 1958).
Elementary text dealing with the Laplace transform, finite Fourier transforms, and, briefly, complex-variable theory. Reasonable mathematical rigor is maintained, but the problems are rather elementary.

COPSON, E. T., *Asymptotic Expansions* (Cambridge: Cambridge Univ. Press, 1965).
Excellent treatment of asymptotic evaluation of integrals.

DITKIN, V. A., and PRUDNIKOV, A. P., *Operational Calculus in Two Variables and its Applications* (New York: Pergamon Press, 1962).

DOETSCH, G., *Theorie und Anwendung der Laplace-Transformation* (New York: Dover, 1943; originally published in Berlin: Springer-Verlag, 1937).
Standard mathematical treatise on the Laplace transform; extensive bibliography of pre-1937 works.

DOETSCH, G., *Handbuch der Laplace-Transformation* (Basel: Birkhäuser, 1950–1956).
Three-volume treatment of theory and applications, both mathematical and physical, of Laplace transform.

ERDÉLYI, A., *Asymptotic Expansions* (New York: Dover, 1956).
Treats asymptotic development of both integrals and solutions to differential equations.

ERDÉLYI, A., *Operational Calculus and Generalized Functions* (New York: Holt, Rinehart, and Winston, 1962).
A brief but authoritative treatment of the modern, rigorous approach to operational calculus; complements Lighthill's (1959) treatment of Fourier integrals.

FOURIER, JOSEPH, *The Analytical Theory of Heat* (Paris, 1822; translated by A. Freeman, Cambridge, 1878; reprinted New York: Dover, 1955).
The fountainhead of Fourier methods.

GARDNER, M. F., and J. L. BARNES, *Transients in Linear Systems* (New York: Wiley, 1942).
Application of Laplace transforms to analysis of lumped-constant, electrical and mechanical systems.

HEAVISIDE, OLIVER, *Electromagnetic Theory* (New York: Dover, 1950 reprint).
Contains much of Heaviside's original use of operational calculus.

LAMB, H., *Hydrodynamics* (Cambridge: Cambridge Univ. Press, 1932; reprint New York: Dover, 1945).
The classical treatise on hydrodynamics.

LIGHTHILL, M. S., *Introduction to Fourier Analysis and Generalised Functions* (Cambridge: Cambridge Univ. Press, 1959).
Deals with the Fourier-transform and -series representations of functions that would conventionally be regarded as improper, and with the asymptotic properties of these representations.

MIKUSIŃSKI, JAN, *Operational Calculus* (New York: Macmillan, 1959).
Original and fundamental treatment of operators in operational calculus [see Erdélyi (1962) for more concise treatment].

MILES, J. W., *The Potential Theory of Unsteady Supersonic Flow* (Cambridge: Cambridge Univ. Press, 1959).
Extensive application of Fourier and Laplace transforms to aerodynamic boundary-value problems.

MORSE, P. M., *Vibration and Sound* (New York: McGraw-Hill, 1948).

SNEDDON, I. N., *Fourier Transforms* (New York: McGraw-Hill, 1951).
Extensive applications of various transforms to physical problems at research-paper level.

THOMSON, W. T., *Laplace Transformation* (Englewood Cliffs, N.J.: Prentice-Hall, 1960).
Similar in scope to Churchill (1958); less rigorous mathematics but more elaborate engineering applications.

TITCHMARSH, E. C., *Introduction to the Theory of Fourier Integrals* (New York: Oxford Univ. Press, 1948).
Standard treatise on Fourier, including Fourier–Bessel or Hankel, integrals and

transforms; largely complementary to Doetsch (1937); extensive bibliography of pre-1948 works.

TRANTER, C. J., *Integral Transforms in Mathematical Physics* (London: Methuen, 1956).
A brief, but not elementary, coverage of all the commonly used transforms.

VAN DER POL, B., and N. BREMMER, *Operational Calculus Based on the Two-sided Laplace Integral* (New York: Cambridge Univ. Press, 1950).
A modern, rigorous presentation of Heaviside's operational calculus as *operational calculus*. Advanced and stimulating applications in such diverse fields as electric circuits and number theory.

VOELKER, D., and G. DOETSCH, *Die Zweidimensionale Laplace-Transformation* (Basel: Birkhäuser, 1950).

WATSON, G. N., *Bessel Functions* (Cambridge: Cambridge Univ. Press, 1945).
The classical treatise on Bessel functions.

Tables and Handbooks

ABRAMOWITZ, M., and I. A. STEGUN, *Handbook of Mathematical Functions* (Washington: National Bureau of Standards, 1964; reprint New York: Dover, 1965).
An extensive tabulation of analytical properties and numerical values of transcendental functions. Highly recommended.

CAMPBELL, G. A., and R. M. FOSTER, *Fourier Integrals for Practical Application* (New York: Wiley, 1948).
The most extensive table of (exponential) Fourier integrals; many entries are effectively Laplace-transform pairs and are presented as such.

ERDÉLYI, A., and J. COSSAR, *Dictionary of Laplace Transforms* (London: Department of Scientific Research and Experiment, Admiralty Computing Service, 1944).
Most of the material from these tables has been included in EMOT.

ERDÉLYI, A., (Ed.), with W. MAGNUS, F. OBERHETTINGER, and F. TRICOMI [EMOT]. *Tables of Integral Transforms*, 2 vols. (New York: McGraw-Hill, 1954).
The most comprehensive tables of integral transforms presently available. Volume 1 contains Fourier-exponential, -cosine, and -sine, Laplace, and Mellin transforms: Vol. 2 contains Hankel transforms, along with many transforms not introduced in the foregoing treatment.

JOLLEY, L. B. W., *Summation of Series* (New York: Dover, 1961).
An extensive tabulation of summable series.

LUKE, Y. L., *Integrals of Bessel Functions* (New York: McGraw-Hill, 1962).
The most extensive table of integrals that contain Bessel functions.

MAGNUS, W., and F. OBERHETTINGER, *Formulas and Theorems for the Special Functions* of Mathematical Physics (New York: Chelsea, 1949).
This valuable (for the applied mathematician) compendium contains short but well-selected tables of both Fourier- and Laplace-transform pairs.

MANGULIS, V., *Handbook of Series for Scientists and Engineers* (New York: Academic Press, 1961).
A tabulation that is both more extensive and more expensive than that of Jolley.

ROBERTS, G. E., and H. KAUFMAN, *Tables of Laplace Transforms* (Philadelphia, Pa.: W. B. Saunders, 1966).
Perhaps the most extensive table of Laplace transforms.

INDEX